MICROBIOLOGICAL ASSAY
An Introduction to Quantitative Principles and Evaluation

MICROBIOLOGICAL ASSAY

An Introduction to Quantitative Principles and Evaluation

WILLIAM HEWITT

Consultant
Quality Control of Pharmaceuticals
Cheltenham, England

formerly
Advisor on Pharmaceutical Quality Control
South East Asia Region
World Health Organization

ACADEMIC PRESS   New York   San Francisco   London   1977

A Subsidiary of Harcourt Brace Jovanovich, Publishers

ACADEMIC PRESS, INC.
111 Fifth Avenue, New York, New York 10003

United Kingdom Edition published by
ACADEMIC PRESS, INC. (LONDON) LTD.
24/28 Oval Road, London NW1

Library of Congress Cataloging in Publication Data

Hewitt, William,   Date
     Microbiological assay.

     Includes bibliographies.
     1.   Microbiological assay.   I.   Title.
QR69.P6H48   576'.028   76-27443
ISBN 0-12-346450-1

148016

CONTENTS

Preface .... *ix*

Acknowledgments .... *xiii*

Chapter 1 INTRODUCTION **1**

1.1 Philosophy of Biological Assay 1
1.2 Basic Techniques and Principles 2
1.3 Mechanization and Automation 5
1.4 Purpose of the Assay 7
1.5 Reference Standards 9
1.6 Preparation of Test Solutions of Standard and Sample 13
1.7 Specifications and Reports 14
     References 16

Chapter 2 THE AGAR DIFFUSION ASSAY **17**

2.1 Introduction 17
2.2 Theory of Zone Formation 17
2.3 Nature of the Response Curve 26
2.4 Dose–Response Curves in Practice 27
2.5 Simple Assay Designs 30
2.6 Simple Multiple Assay Designs 36
2.7 Simplification of Computation of the Potency Ratio 38
2.8 Designs Incorporating Checks for Curvature (Three Dose Levels) ... 39
2.9 Designs Incorporating Checks for Curvature (Four Dose Levels) 43
2.10 Assays by Large Plates—General Principles 48
2.11 Large Plate Assays Using Latin Square Designs 52
2.12 Large Plate Assays Using Quasi-Latin Square Designs 55
2.13 Low-Precision Assays Using Large Plates 57
2.14 Small Plate Assays Using Interpolation from a Standard Curve 61
2.15 Missing Values 66
     References 68

Chapter 3 TUBE ASSAYS FOR GROWTH-PROMOTING
              SUBSTANCES **70**

3.1 General Principles 70
3.2 Measurement of Response 72
3.3 Nonideal Responses 74

3.4   Slope Ratio Assays—An Unbalanced Design 81
3.5   Criteria of Validity 83
3.6   Multiple Linear Regression Equations 85
3.7   Potency Computation from an Unbalanced Design 87
3.8   Balanced Slope Ratio Assays with Linearity Check 90
3.9   Simplified Computation of Potency Ratio from Balanced Assays 93
3.10 Multiple Assays by the Slope Ratio Method 97
      References 101

Chapter 4   TUBE ASSAYS FOR ANTIBIOTICS **103**

4.1   General Principles 103
4.2   Response Curve—Commonly Used Forms of Expression 105
4.3   Linearization of Dose–Response Relationships—Theoretical Considerations .. 107
4.4   Dose–Response Linearization Procedures in Practice 113
4.5   Potency Estimation by Interpolation from a Standard Curve 123
4.6   Graphic Estimation of Potency by Probit of Response versus Dose 125
4.7   Arithmetical Estimation of Potency from an Assay of Balanced Design by
      Angular Transformation of Response 127
4.8   Graphic Estimation of Potency Using the Relationship Log Response versus
      Dose 130
4.9   Estimation of Potency by Interpolation from a Dose–Response Line Using
      an Automated System 132
      References 135

Chapter 5   ASSAY OF MIXTURES OF ANTIBIOTICS **136**

5.1   Occurrence of Mixtures and the Nature of the Problem 136
5.2   General Techniques 139
5.3   Comparative Bioautographs 142
5.4   Quantitative Bioautographs 145
5.5   Differential Assays 147
      References 150

Chapter 6   EVALUATION OF PARALLEL LINE ASSAYS **151**

6.1   Introduction 151
6.2   Basic Assumptions for the Statistical Evaluation of Parallel Line Assays 154
6.3   Evaluation of a Standard Log Dose–Response Curve 154
6.4   Analysis of Variance to Separate Components Attributable to Various Sources 156
6.5   Evaluation of a Simple Two Dose Level Assay 162
6.6   An Alternative Method for Obtaining Confidence Limits 167
6.7   Evaluation of a Multiple Two Dose Level Assay 167
6.8   Evaluation of a Three Dose Level Assay 170
6.9   Evaluation of a Large Plate Assay (Latin Square Design) 173
6.10 Evaluation of a Large Plate Assay (Quasi-Latin Square Design) 177
6.11 Evaluation of a Large Plate, Low Precision Assay 183
6.12 Evaluation of an Assay Incorporating Reference Points—The "FDA" Design   187

6.13   Evaluation of a Quantitative Bioautograph   191
6.14   Evaluation of Tube Assays Using Function of Response versus Logarithm of
       Dose   194
       References   202

Chapter 7   EVALUATION OF SLOPE RATIO ASSAYS   203

7.1   Principles of Evaluation   203
7.2   Evaluation of Simple Slope Ratio Assays   204
7.3   Evaluation of a Multiple Slope Ratio Assay   207
7.4   Evaluation of a Turbidimetric Antibiotic Assay by Angular Transformation of
      Response and Slope Ratio   210
      References   213

Chapter 8   CHOICE OF METHOD AND DESIGN   214

8.1   Choice of Method   214
8.2   General Considerations in Selection of Design   216
8.3   Plate Assay Designs   217
8.4   The Influence of Curvature in Parallel Line Assays   225
8.5   Choice of Design for Plate Assay   228
8.6   Tube Assays—General Considerations   232
8.7   Designs for Slope Ratio Assays   233
8.8   Influence of Curvature in Slope Ratio Assays   238
8.9   Choice of Design for Slope Ratio Assays   240
8.10   Tube Assays for Antibiotics   241
       References   242

Chapter 9   REPEATED ASSAYS, SPECIFICATIONS, AND REPORTS .   243

9.1   Replication of Assays   243
9.2   Collaborative Assays   245
9.3   Combination of Replicate Potency Estimates   247
9.4   Combination of Replicate Potency Estimates—Simplified Methods   251
9.5   Specifications for Antibiotics   253
9.6   Official Standards—A Practical Approach   254
      References   257

Appendix 1   PATTERNS FOR SMALL PLATE ASSAYS   259

Appendix 2   CALCULATIONS FOR PARALLEL LINE ASSAYS   260

Appendix 3   POTENCY RATIO, OR *F/E*, TABLES..................   262

**Appendix 4   PRO FORMA AND WORKED EXAMPLE—
                AGAR DIFFUSION ASSAY 264**

Appendix 5   SOURCES OF REFERENCE MATERIALS **265**

A5.1   International Biological Standards and International Biological Reference
Preparations   265
A5.2   British Biological Standards and Reference Preparations   265
A5.3   European Pharmacopeia Commission Reference Substances   265
A5.4   United States Pharmacopeia Reference Substances   265
A5.5   International Chemical Reference Substances   266

Appendix 6   THE DILUTION OF REFERENCE STANDARDS **267**

Appendix 7   THE PROBIT TRANSFORMATION **269**

Appendix 8   THE ANGULAR TRANSFORMATION **270**

Appendix 9   THE t DISTRIBUTION **271**

Appendix 10   VARIANCE RATIO TABLES—THE F TEST **272**

Appendix 11   THE χ^2 DISTRIBUTION **274**

Appendix 12   THE RANGE/MEAN TEST **275**

Appendix 13   EVIDENCE THAT QUADRATIC CURVATURE IS
WITHOUT INFLUENCE IN BALANCED
PARALLEL LINE ASSAYS **276**

Index   279

PREFACE

During the past decade there has been world-wide increased awareness of the importance of quality control of pharmaceutical preparations. The World Health Organization has at the same time been active in encouraging and assisting member states to strengthen their national quality control systems. As a result, many nations are planning either new regulatory laboratories or the extension of existing facilities.

The increased importance of microbiological assay of antibiotics is indicated by the existence today of about forty International Biological Standards and Reference Preparations as compared with one in 1948.

Quality control of this large and important group of modern medicines is a major task of pharmaceutical analysts both in industry and the laboratories of government regulatory authorities. Although physical and chemical methods of examination may in some cases suffice, in many cases microbiological assay remains the only method of assessment of potency. The value of microbiological assay for the estimation of certain vitamins and amino acids is also well established.

The method has the advantage that it *can* be carried out without highly specialized and expensive equipment. There have been developments in mechanization and automation of individual steps in assay procedures. Completely automated systems are also available. These call for relatively large capital expenditure which can be justified only when the examination of a sufficiently large number of samples is envisaged. It seems likely that manual methods will continue to be widely used for many years.

As the same basic procedures are applicable to a wide range of antibiotics, microbiological assay is well suited for the routine examination of large numbers of samples by well-trained and well-supervised personnel. The full potential of the method is achieved in those laboratories in which it is treated as a branch of quantitative pharmaceutical analysis and applied with an awareness of the chemical, physical, and mathematical as well as biological principles involved.

Often, however, the method is applied empirically with inappropriate and inefficient designs, poor reproducibility being attributed to "biological error."

While random variation of individual responses is a significant feature of most microbiological assays, there seems to be no reason to suppose that this is biological in origin. The concept of biological variation is based on the differing responses of individual subjects to the same stimulus. In *macro*biological assays where the number of test subjects receiving the same stimulus is small, random variation of mean responses is to be expected. In *micro*biological assay, the situation is quite different. Inocula are in most cases measureable in terms of millions of organisms so that mean responses to the same stimulus under the same test conditions are likely to be very similar. The problem facing the analyst is not uncontrollable biological variation of the organism but how to regulate the physical conditions of the test so as to ensure that the effect of the stimulus is not modified by unwanted influences.

Despite such control of physical conditions, as in any assay method, random variation of responses remains and may be measured by statistical techniques.

In routine *micro*biological assay it is in the author's opinion neither necessary nor economically justifiable to carry out a statistical evaluation leading to confidence limits for each individual assay. Moreover, routine calculation of such limits can be a trap for the unwary, giving false confidence in estimated potencies when bias due to poor techniques may be substantial yet overlooked.

It is for these reasons that statistical evaluation of assays is treated in Chapters 6 and 7 quite separately from the potency calculations of Chapters 2–4.

The need to be aware of the principles of statistical evaluation becomes apparent in Chapter 8 in which features of assay design such as replication, number of dose levels, and spacing of dose levels are discussed.

Many supervisors of microbiological assay laboratories are by inclination bacteriologists rather than analysts; thus, the mathematics has been kept as simple as possible. It is assumed only that the reader has an elementary knowledge of algebra and has been introduced to the basic concepts of statistics. Although some calculations are lengthy, the individual steps are nothing more than simple arithmetic. Practice in the application of these methods helps the beginner gain a better understanding of their principles. A word of warning: statistical evaluation as normally applied to simple individual assays gives an idea of the capability of the method, or to put it another way, it gives an idea of the limitations imposed by technique and assay design. It does not take into account the gross errors or biases which can and do arise from neglect of the special features of the assay method and the principles of quantitative analysis. Thus, evaluation of *precision* is no substitute for painstaking efforts to control physical conditions and operating procedures so as to obtain an *accurate* estimate of potency.

The examples given in this work have been collected from several laboratories in North America, Europe, and Asia over a period of more than fifteen years. Many represent routine work of these laboratories and do not necessarily indicate the author's ideas on either analytical technique or assay design. The examples have been chosen using the criterion that they illustrate a representative selection of designs available to the microbiological analyst.

The use of poor assay design is widespread. Such designs have been included here very deliberately so as to draw attention to their disadvantages.

This volume has developed from notes written in Turkey in 1968 with the limited aim of explaining the mysteries of the calculation procedures employed in microbiological assay. Although this remains an important aspect of the work, there is now stress on assay design, the elementary principles on which the various methods are based, as well as general principles of pharmaceutical analysis.

No attempt is made to give detailed descriptions of individual assay methods. This information is available from sources such as the international and various national pharmacopeias, the United States Code of Federal Regulations, Kavanagh's "Analytical Microbiology" Volumes I and II, Barton-Wright's "Microbiological Assay of the Vitamin-B Complex and Amino Acids," and György's "Vitamin Methods" Volumes I and II.

More advanced accounts of principles of microbiological assay are to be found in certain chapters of Kavanagh's "Analytical Microbiology" Volumes I and II. For advanced accounts of assay design and evaluation the reader may consult Finney's "Statistical Method in Biological Assay" or Bliss's contribution to György's "Vitamin Methods" Volume II.

It is hoped that this work will succeed in its aim of providing an elementary introduction to the principles of microbiological assay, assay design, and calculation procedures and so help to encourage a less empirical approach to the subject.

ACKNOWLEDGMENTS

I should like to acknowledge the part played directly or indirectly by many persons in making this work possible. Mr. D. F. Harris, Mr. S. Pugh, and Mr. J. S. Simpson who introduced me to the practice of microbiological assay; Mr. J. P. R. Tootill who first inspired my interest in assay design; Mrs. Ülku Güngör, Mrs. Meliha İnak, Miss Ülkü Önal, Mrs. Secil Sade, and Mrs. Waralee Sithipitaks for their very willing cooperation in practical assay procedures; Miss M. A. Garth, Miss A. Jones, Dr. J. W. Lightbown, Mr. G. A. Stoddart, and Dr. W. W. Wright for informal discussions which have kept me informed of their experience especially with reference to automated methods; Dr. V. N. Murty and Mr. J. R. Murphy for advice on certain statistical problems; Dr. K. Tsuji for correspondence on curve straightening procedures; and finally Dr. F. W. Kavanagh for his many suggestions, constant interest, and enthusiasm for the subject which has influenced the work greatly.

It is natural that the end product does not necessarily reflect accurately the views of everyone whose help is acknowledged and that any shortcomings are of course my own responsibility.

I wish also to thank Dr. H. Mahler, Director General of the World Health Organization, for permission to publish this work.

CHAPTER 1

INTRODUCTION

1.1 Philosophy of Biological Assay

The need for standardization of products affecting our lives in literally thousands of ways is a well-established fact. It was expressed picturesquely in advertisements for a certain brand of shaving soap, which was claimed to produce just the right amount of lather: "Not too little, not too much, but just right!"

The need for standardization assumes much greater importance where medicinal substances are concerned. In some cases the margin between too little (an ineffective dose), and too much (a toxic dose) may be relatively small. In other cases, while an unnecessarily high dose may not be toxic it could be undesirable on economic grounds.

Many medicinal agents consist of a single active substance that can be characterized completely in terms of its chemical, physicochemical and purely physical properties. A specification may be devised for a pharmaceutical grade taking into consideration the properties of the pure substance and making allowance for tolerable levels of impurities. Certain impurities that are expected to arise from the manufacturing process may be limited by specific tests.

Other medicinal agents, however, particularly those of natural origin, may be of more variable character. They may consist of a mixture of chemically related substances differing quantitatively and qualitatively in their biological effects. They may also include chemically unrelated substances that have biological activity. The activities of the different components may be either mutually compatible or antagonistic. They may even be synergistic.

Not uncommonly, a substance becomes of recognized therapeutic value before its exact chemical composition has been ascertained.

Such problems existed long before the discovery and commercial production of antibiotics with which we are largely concerned in this present work. Well-known examples include the alkaloids of ergot, the solanaceous alkaloids, digitalis glycosides, and the purgative drugs containing anthraquinones and related substances.

When for any reason a potentially valuable medicinal agent cannot be defined in terms of its chemical or physicochemical properties, then the obvious alternative is to consider its biological properties.

Unfortunately, biological properties cannot be simply quantified. Attempts to measure potencies by purely biological means have never been successful due to the inherent variability of the biological system. It is true that some tests such as those described for pyrogens and toxicity in certain pharmacopeias are defined in terms of the effect of the drug on a group of animals under specified conditions. However, these are limit tests. Moreover, it is likely that results would not be closely reproducible in different laboratories or even in the same laboratory on different occasions. The limitations of these methods are recognized and accepted in the absence of better alternatives.

For truly quantitative work the problem of variable response of the test organism is overcome by the use of comparative methods. A quantity of active substance is set aside and designated the standard preparation. The effect of any sample on a biological system can be compared with this standard preparation to obtain a quantitative relative potency.

The method used for comparison of the two preparations may be macrobiological, such as the assay of insulin using mice, or microbiological, such as the assay of streptomycin or tetracycline.

1.2   Basic Techniques and Principles

The two most commonly used methods of microbiological assay will be referred to as the plate (or agar diffusion) method and the tube method. The basis of both these methods is the quantitative comparison of the effect of two substances on the growth of a suitable microorganism in a nutrient medium. The two substances are a standard and a sample whose potency is to be determined. The effect may be to inhibit growth, as in the case of antibiotics, or to promote growth, as in the case of vitamins and amino acids.

The practical procedures for both these methods are illustrated here only in outline by two typical simple antibiotic assays.

(a)   *The plate assay of penicillin.*   Nutrient agar is melted and its temperature reduced to 48°C. A small volume of a suspension of a penicillin-sensitive microorganism (e.g., *Staphylococcus aureus*) is added by pipet and gently but well mixed to give a uniform dispersion in the agar medium. A suitable volume (about 15–20 ml) of this seeded agar is pipetted into a petri dish to give a layer of uniform thickness (about 3–5 mm).

After solidification of the seeded agar the plate is ready for use and may be refrigerated until required.

Two or more concentrations of penicillin solutions prepared from both reference standard and test sample are applied to reservoirs at appropriately spaced positions on the plate. These positions may be in accordance with different randomized patterns for each plate in a set comprising one assay.

One experimental design, however, uses only a single pattern for all plates (see Example 8).

Suitable patterns for use with petri dishes are given in Appendix 1.
Various forms of reservoirs are in use:

(1)   Small cylinders of the agar are cut and removed using a cork borer, or better, a specially made 8-mm diam stainless steel punch is convenient.

(2)   Specially designed sterilized stainless steel cylinders are placed on the surface of the agar.

(3)   Small sterilized electrical ceramic insulators (fish spine beads) are used. The beads are dipped into the test solution, surplus liquid is drained off, and then the bead is placed on the agar surface.

(4)   Small filter paper disks are used in a similar manner to the beads.

For the first two of these procedures a standard volume of test solution is added to each reservoir. This may be measured simply as a constant number of drops added from a standard dropper, or the reservoirs may be filled almost to the brim, or a semiautomatic pipette may be used.

The solution is allowed to diffuse into the agar at room temperature or lower for an hour or perhaps more, and then the plates are incubated, usually overnight.

After incubation, clear zones surround the point of application of the antibiotic, whereas in other parts of the plate growth of the microorganism causes turbidity.

Zone boundaries are usually clearly defined, although the sharpness of definition varies according to test organism, the density of the inoculum, the antibiotic, etc.

Inhibition zone diameters are measured. The relationship between mean responses (zone diameters) to each test solution and the concentration of that test solution is the quantitative basis of the assay.

(b)   *The tube assay of neomycin.*   A series of concentrations of neomycin standard solutions are prepared, as well as one or more solutions of the sample within the same concentration range as the standard. For one series of tubes, 1 ml of each solution is added to a separate test tube. This is followed by 9 ml of a nutrient medium inoculated with a suspension of the neomycin-sensitive test organism *Klebsiella pneumoniae*. Usually two or more series of tubes are included in each assay. The tubes are incubated for about 4 hours; then the growth is stopped in all tubes at the same time (immersion in a water bath at 80°C is very satisfactory—higher temperatures may result in coagulation of protein). The growth of the organism is estimated by the turbidity measured in a suitable photometer.

The mean inhibition of growth corresponding to each test solution in the set is the basis of calculation of potencies. Lower turbidities correspond to higher concentrations of antibiotic.

Despite their widely differing techniques, these two methods have a common basis in that they depend on the following principles:

(1) Comparison of a sample of unknown potency with a standard substance of known defined activity.

(2) Measurement of the inhibiting effect on the multiplication of the test organism.

(3) The existence of some form of quantitative relationship between concentration of active substance and response.

(4) This quantitative relationship is the same for the sample as for the standard.

The forms of these relationships and convenient ways of calculating potencies are described in detail in Chapter 2 for agar diffusion assays and Chapters 3 and 4 for tube assays.

Both plate and tube assays of growth-promoting substances (g.p.s's) such as vitamins and amino acids differ from assays of growth-inhibiting substances in that the response is opposite. In plate assays the point of application of the test solution is surrounded by a turbid zone of exhibition contrasting with its relatively clear surroundings. In the tube assay increasing doses of test solution cause increasing growth of the organism.

Both techniques have the following requirements:

(1) The test organism must be dependent for growth on the presence of the substance to be assayed.

(2) Addition of graded doses of the substance to be assayed (both sample and standard) should result in graded responses on incubation.

(3) The nutrient medium for the test must contain an excess of all substances required by the test organism except the substance to be assayed. This substance should be absent from the basic medium.

(4) Apart from the substance to be determined, no other substance that may be present in the sample should be capable of promoting growth of the test organism or of modifying its growth. This is a factor to be considered in choosing the test organism. It is an ideal that is sometimes difficult to attain.

For assays of g.p.s.'s, in contrast to the assay media for antibiotics, etc., a synthetic medium must be devised so as to ensure compliance with requirement (3). This medium may include, for example, buffers, vitamin-free casein, glucose plus traces of amino acids, and vitamins other than the g.p.s.'s to be estimated.

A typical plate assay method is that for cyanocobalamin using *Escherichia coli*. This assay is used to illustrate a large plate quasi-Latin square design procedure in Example 6.

Tube assays for g.p.s.'s are the subject of Chapter 3.

A principle of both antibiotic and g.p.s. assays is described by Jerne and Wood (1949) as the "condition of similarity." That is to say, if the substance in the standard preparation that causes the characteristic response in the test subject is described as the *effective constituent*, then the response to the "unknown" test preparation must also be due only to the same effective constituent and be unmodified by other substances. In other words, the less potent of the two preparations (standard and test) that are being compared behaves as though it were a dilution of the other in an inert diluent.

It follows that when this principle is observed a change in experimental conditions, test organism, or response measured will not influence the true potency ratio between the two preparations. Any differences in estimated potency ratio would be attributable to experimental or random error only.

It also follows that in designing any assay procedure, it is necessary to take into account the possible influence of substances other than the effective constituent that may be present in either of the preparations. This is discussed further in Sections 1.5 and 1.6 and in Chapters 2–5, 8, and 9.

When the condition of similarity is truly applicable then the choice of test organism is dependent only on practical convenience, e.g., adequate sensitivity, sharpness of zone boundaries in plate assays, and slope of the response line. In practice, while very often the standard reference preparation may approach the ideal of being a dilution of the effective constituent in an inert diluent, the same may not be true of the test preparation. The latter may contain other active substances, either naturally occurring or as admixtures in pharmaceutical formulations. In such cases, the use of a test organism that is insensitive to the additional active constituent is necessary. Some examples are given in Chapter 5.

The practical importance of using the correct culture is clear.

1.3 Mechanization and Automation

Over a period of many years means have been sought to improve reliability and increase output of assay results. Mechanical aids have been developed for certain operations of manual assays and in more recent years partially and fully automated methods have been developed.

Mechanical aids include automatic diluters and media dispensers for liquid broth and molten agar. Coffey and Kuzel (1966) devised a machine for pouring the thin inoculated upper layer in double-layer petri dishes. An accurately measured volume of molten medium is applied to the solid base layer by means of a nozzle, which sweeps rapidly above the plate in a spiral path, thus ensuring a layer of uniform depth. The importance of this uniformity is discussed in Section 2.2.

Application of test solutions to reservoirs may be made by means of a semiautomatic pipet such as the Oxford® SAMPLER Micro-Pipetting System (Oxford Laboratories, San Mateo, California). This system uses a single pipet with separate tips for each test solution. It is so designed that there is no carry-over of test solutions. It has been found very convenient in large plate assays in which the filling order of reservoirs is strictly defined (Stoddart, 1972); see Sections 2.10–2.12.

The Fisher–Lilly Antibiotic Zone Reader (Fisher Scientific Company, Pittsburgh, Pennsylvania) is a very widely used aid for measuring zone diameters. This is a compact unit that projects a magnified image of the zone onto a screen that is an integral part of the unit. The image is positioned so that a hairline on the screen meets one zone edge tangentially. The plate is then caused to move in a straight line by rotation of a calibrated drum until the opposite zone edge coincides with the hairline. The distance moved is read in millimeters from the drum, which is graduated at 0.2-mm intervals. Readings are easily estimated to the nearest 0.1 mm. This equipment is designed for use with petri dishes only and cannot be used with the large plates that are described in Section 2.10.

The Biocoder System (R. N. Saxby Ltd., Liverpool) uses a projection system and a caliper that is adjusted by the operator to measure the projected zone image. The caliper has an electrical readout that is signaled to the Biocoder on depression of a foot switch by the operator. The Biocoder records and processes the signal.

Completely automated reading of plates is provided by such systems as the Auto Biocoder (R. N. Saxby Ltd., Liverpool), the Autodata (Autodata, Hitchin, Hertsfordshire, England), and the Quantimet® 720 (IMANCO, Melbourn Royston, Hertsfordshire, England). All three of these systems have provision for detection of spurious zones, i.e., zones that are imperfectly shaped or are grossly over- or undersized. These systems have the advantage of providing completely objective measurements at high speed.

An almost completely automated turbidimetric assay system is exemplified by the Autoturb® (Elanco Products Company, Indianapolis, Indiana). This incorporates the following functions:

(1) Automatic dilution of the test sample to different dose levels with broth and dipensing into tubes held in racks.

(2)   Manual transfer of the racks of tubes to a high-precision, temperature-controlled incubation bath.

(3)   At the end of the incubation period (e.g., after 4 hours), manual transfer of the racks to an 80°C water bath to kill the organism thence via a cooling bath to the reading module.

(4)   Automatic sampling and reading of the optical transmittance of the incubated broth from each tube. The photometer employs a flow-through optical cell that ensures high reproducibility of measurements even in the case of rod-shaped organisms. The problems of measurement of optical properties of cell suspensions are discussed in Section 3.2.

(5)   Automatic recording of responses and computation of sample potencies.

This equipment may be used for both antibiotic and vitamin assays. Accuracy comparable with that of chemical methods is claimed and it appears that output of results is limited normally only by the capability of the operator to prepare samples for presentation to the sampling and diluting module. This module will accept over 100 samples per hour and the reading module over 75 samples per hour.

The Autoturb ® and its operations and performance are described in a series of articles by its designers (Kuzel and Kavanagh, 1971a,b; Kavanagh, 1971, 1974).

The Technicon AutoAnalyser (Technicon Instruments Surrey, England) is a very versatile system, which was developed for completely automated colorimetric chemical assay methods. Its adaptation for microbiological assay is described by Haney *et al.* (1962), Gerke *et al.* (1962), Shaw and Duncombe (1963), Jones and Palmer (1970), and Grimshaw and Jones (1970).

It is claimed that as compared with the manual plate method, both precision and output of results are increased. However, the system does not appear to be widely used in microbiological assaying.

1.4   Purpose of the Assay

While the need to determine potency may almost always be ultimately related to health services, the immediate purposes of particular assays may vary greatly. The type and design of assay used may vary according to this purpose. In research and development on new substances, samples ranging through fermentation broths, crude extracts, and partially purified substances may be assayed using several test organisms. Chromatographic techniques such as the bioautograph (see Chapter 5) may supplement the conventional potency comparisons. As will be described later (Chapter 5) discrepancies between assays using different test organisms may yield important clues as to the varying composition of mixed antibiotics.

Economic manufacture of antibiotics and some vitamins that are made by a fermentation process necessitates process control procedures, which often include microbiological assay. Active substance levels are relatively high and so high sensitivity is not a requirement of the assay. Speed, however, is probably important. A fermentation process may take only two or three days. If the harvest time is to be decided on the basis of an assay, then clearly that assay method must be capable of yielding a result within a few hours.

In contrast, studies of the absorption and excretion rate of an antibiotic by animals or humans may need more sensitive methods because of the lower antibiotic levels in the samples. Methods that are both sensitive and rapid are needed in assays for relatively toxic antibiotics such as gentamycin in body fluids during the course of clinical treatment so that dosage can be controlled accurately.

For assessment of the quality of the finished refined active substance, neither speed nor sensitivity is important. A reliable estimate of potency, however, is necessary and so a precise and bias-free assay procedure must be employed. Similar considerations apply in the batch control and long-term stability testing of pharmaceutical dosage forms.

Independent control of products offered for distribution to the public often necessitates the planned sampling of products from many manufacturers. The same substance may be offered in a variety of dosage forms. Products may have been subjected to storage conditions differing greatly in both duration and severity. For a comprehensive survey of products on the market then, the independent analyst needs to organize his work in such a way as to deal with a large number of similar samples. There is no requirement for his methods to be either rapid or sensitive. The precision required for this type of work is not necessarily always high. This will be discussed in Chapters 8 and 9.

It should be noted that while in some cases chemical or physicochemical assay may be perfectly reliable for product quality control in industry, it may not be suitable for either stability testing by the manufacturer or independent control by a public authority. In the routine production batch control of chloramphenicol dosage forms, for example, provided that the raw ingredient chloramphenicol conforms to specification and the process does not include any steps that could possibly cause deterioration, then the freshly made batch may be checked for strength by means of its UV absorption character-istics. In an old and deteriorated product, however, drop in potency as determined by microbiological assay is not paralleled by a drop in UV absorption. Thus a direct UV absorption measurement is unsuitable.

1.5   Reference Standards

As the basis of any microbiological assay is the comparison of a sample with a reference standard, it is clear that the standard is of fundamental importance. Any differences between standards used in different laboratories or on successive occasions in the same laboratory will lead to changing biasses in the estimated potency of a single sample.

Ideally, standard reference material should be:

(1)   Available in sufficient quantities for all assays over a long period.
(2)   Completely homogeneous.
(3)   Stable.
(4)   Qualitatively identical with substances to be tested.
(5)   Preferably a standard should be a single substance, but if it is a mixture, then the various components should be present in the same proportions as the product to be tested.

Miles (1952), Lightbown (1961), and Wright (1971) give excellent expositions on the basis of biological standardization with particular emphasis on reference standards. These are articles well worthy of study by all analysts involved in biological testing and also by the recipients of results of biological assays.

Miles's illustrated account of the development of a hypothetical standard is reproduced here by kind permission of the Director General of the World Health Organization. (The ideal requirements numbered (4) and (5) correspond to Miles's "hypotheses of similarity.")

> In [assays] where the standard and test preparations are substantially pure, the conditions for valid assay are almost certainly fulfilled. In the assay of antibiotics in general, however, we have no right to assume they will be fulfilled, because either of the preparations may be impure. Suppose that the first usable crude preparation of an antibiotic is made into a standard and, although we do not know it, it contains, besides inert impurities, two distinct molecular species (α and β) of the antibiotic—as distinct say as streptomycin and mannosido-streptomycin. Suppose also that the crude preparations assayed against this first standard contain only these two members of this particular family of antibiotics (Fig. 1.1). If all the test preparations were like B, and contained the same proportions of α and β, the hypothesis of similarity holds, and the assay in terms of A will be valid. But the likelihood of preparations like D and F and even more complex mixtures makes any confident assumption of the hypotheses of similarity impossible. Both standard and test preparations are in fact likely to be heterogeneous.
>
> With the next advance in purification it may be possible to produce a standard consisting of one molecular species, either pure or with some inert matter (C), but it may still be necessary to use it in the assay of heterogeneous preparations such as B or D.
>
> In fact we do not reach the valid assay stage until we know without question that the standard is like E and test preparations are like F.

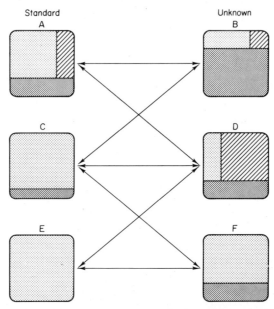

Standard
A

Unknown
B

C

D

E

F

Fig. 1.1. A diagrammatic representation of possible changing composition of reference standard and "unknown" samples during the development of a new antibiotic. Standard is represented by A, C, and E, and "unknown" preparations are represented by B, D, and F. Light area: molecular species α; hatched area: molecular species β; dark area: inert impurity. (Adapted from Miles, 1952.)

In practice of course, the ideals are not always attainable and so a compromise must be sought.

During the early stages of study of a new active substance the laboratory concerned will usually set aside a quantity of fairly typical material sufficient for its own use during the development period. The material is arbitrarily assigned a potency that is usually expressed in terms of units of activity per milligram. That is to say, the unit is defined as the activity present in an arbitrarily specified weight of this particular reference material.

As supplies of this original reference material diminish it becomes necessary to establish a new standard preparation. Often more highly purified and therefore more potent materials are available by this time. A quantity is set aside for use as a new reference material. The potency assigned to this may be in terms of the original standard and determined by repeated assay using the old substance as standard and preferably both plate and tube assays and more than one test organism. The problems that may be encountered in such a test are discussed in Chapter 9. Having established the relative potency of the new reference material, the unit may be redefined in terms of the weight of the new material that contains one unit of activity.

Many antibiotics are important items of international commerce and as such it is necessary that the potency of a single antibiotic, whatever its country of origin, be definable in terms of a common standard.

The World Health Organization has been instrumental in promoting the establishment of International Biological Standards, and International Biological Reference Preparations will be described here collectively as "international reference materials." The decision to establish such a material is taken by an Expert Committee of the WHO, which assesses worldwide need. Material is selected by the committee taking into consideration the views of interested parties. The views of various primary manufacturers who have carried out research and development work are of great value in ensuring that a realistic and universally acceptable reference material is selected.

International standards are usually established after international collaborative tests to confirm the suitability of the selected material. A potency is assigned in terms of International Units per milligram (IU/mg).

International collaborative assays involve a massive effort both in organization and interpretation of results.

International Reference Preparations may be established without subjecting the selected material to such rigorous studies and do not always have a potency assigned to them in terms of IU/mg.

In the Supplement (1971) to the second edition of the International Pharmacopoeia (1967), potencies have been assigned to paromomycin, cefalotin, and lincomycin, which had previously been issued as Reference Preparations without potencies.

An International Biological Standard or International Biological Reference Preparation is intended to serve as a fundamental standard to be observed by all countries. A typical individual material may consist of about 1 kg packed into ampuls each containing about 50–150 mg. Thus the quantity available for distribution to any individual laboratory is strictly limited. The intention is that each country should set up its own national standards, calibrating them against the international materials.

Although there is widespread distribution of the international materials, it seems that few countries have formally established national standards in this way.

For routine use each laboratory concerned with antibiotic potency testing needs for each antibiotic a "working standard" that is available in adequate quantities and whose potency has been accurately determined relative to an "official standard." Precise measurement of potency as needed for a standard may be a laborious business (see Chapter 8). A collaborative assay involving several laboratories reduces the workload on individual laboratories and also has the advantage of minimizing any bias that may be a feature of the work of a single laboratory. However, collaborative assays are not without their problems (see Chapter 9).

The concept of a unit of biological activity defined as the specific activity contained in a certain weight of a unique standard specimen of the material existed many years before the advent of antibiotics.

The term "specific activity" is noteworthy. The unit of activity, a characteristic of the particular substance, is quite arbitrary and should not be interpreted to mean that the numerical values of potencies of different antibiotic substances give any indication of their relative activities. Relative activities, as measured for example, by minimum inhibitory concentration, vary according to the test organism. Potency of a standard, however, in terms of units per milligram is fixed by definition regardless of the test organism.

Lightbown (1961) draws attention to the confusion that has resulted from the introduction of the concept of "microgram equivalent." The intention was to express the unit as being the activity of one microgramme of the active substance. In many cases the international unit is identical with the "microgram equivalent" of activity. However, the use of this terminology has lead to such anomalies as the potency of tetracycline base being 1082 μg/mg.

Sometimes the need to examine pharmaceutical dosage forms arises before any "official" standard has been set up. In such cases a compromise solution must be found.

A company wishing to control its pharmaceutical processing or to carry out stability tests may find it convenient to use the ingredient material as reference standard.

A laboratory of a regulatory authority may be able to compare the physical, chemical, and biological properties of the ingredient material from two or more sources with one another as well as with published figures. Such a study could lead to the setting up of a provisional laboratory standard.

When a substance can be characterized adequately by chemical and physical means, there is no need for a biological standard. It is for this reason that the International Reference Standard for benzylpenicillin is to be discontinued when present stocks are exhausted.

Even when the antibiotic itself can be assessed by chemical methods, microbiological assay often remains the most convenient procedure for the examination of pharmaceutical and clinical samples, etc. For these assays, the reference standard may be a substance of known identity and purity as determined by chemical and or physical means.

For the semisynthetic penicillins, no biological reference material is needed. International Chemical Reference Substances are available, however, and the International Pharmacopeia prescribes the use of the ampicillin reference substance for the microbiological assay of the various forms of ampicillin and its pharmaceutical preparations. The British Pharmacopeia relies entirely on chemical assay for the assessment of purity of ampicillin and control of its preparations.

In contrast to the biological reference materials, International Chemical Reference Substances are available in greater quantities and are intended as working standards. Some sources of reference standards are listed in Appendix 5.

1.6   Preparation of Test Solutions of Standard and Sample

The problems encountered in sample preparation have much in common with those of chemical analysis. If the sample to be examined is the active substance itself, then the method of preparation of the sample test solution is normally identical with that for the standard reference material. If, however, the material to be examined is a crude or partially purified natural product, a clinical sample, or a compounded pharmaceutical preparation, then some process may be necessary to separate quantitatively the active substance from potentially interfering substances or to make allowance for their presence.

In devising an assay procedure the questions that must be considered include:

(a)   If other active substances are present, will they have an effect on the particular test organism under the proposed test conditions? These problems are discussed in Chapter 5 for antibiotics.

(b)   Will "inert" substances that are present modify the response? Such influences might be

(1)   adsorption of the active substance on a solid phase,
(2)   partition of the active substance between two liquid phases,
(3)   influence of salts in the diffusion assay,
(4)   possible enrichment of the medium by sugars, etc.,
(5)   binding of the active substance with protein as in samples of blood or milk.

Some specific extraction techniques are given by Grove and Randall (1955).

Having decided on technique, then it is necessary to devise a quantitative method of test solution preparation involving realistic weighings and volumetric measurements. The following simple principles should be borne in mind:

(1)   *Weighings.*   Too small a weight of sample may introduce a weighing error, or if the sample is not homogeneous, a sampling error; too large a weighing may necessitate excessive dilution, thus introducing additional dilution errors.

(2)   *Adjustment for moisture content.*   It is frequently required that potency of an antibiotic substance be expressed in terms of the dry material. The apparently simple process of drying the material in an oven is very prone to

error. Streptomycin sulfate, for instance, holds its moisture content tenaciously. After drying, it will rapidly regain moisture unless suitable precautions are taken. If the quantity of material is limited as in the case of reference materials, then the actual weight loss may be very small, for example, 4 mg loss in 100 mg of material. The loss in weight of adsorped moisture from an "air dry" weighing bottle may be as much as 10 mg. It is clear that unless this adsorped moisture is completely removed then substantial errors are likely. The following case illustrates the seriousness of the problem. In a collaborative assay, seven laboratories determined the moisture content of vials containing 250 mg of streptomycin sulfate. Despite detailed written advice on the procedure for measuring loss on drying at 60°C in vacuum, results reported were spread fairly evenly between 1 and 5%.

Further investigations revealed no evidence of substantial variation from vial to vial and the moisture content was established as being in the range 4.1–4.6%. Clearly incorrect determination of moisture content of sample or standard can be a major source of error in microbiological assay.

(3) *Volumetric measurements.* Small volumes such as 1 or 3 ml and fractional measurments such as 6.4 ml should be avoided, since these contribute appreciable errors; glassware manufacturer's tolerances increase with decreasing size of pipet; almost invariably suitable dilution schemes can be devised using no volume measurement lower than 5 ml; there is no need whatsoever to work with exact "round" figures. See also Appendix 6, which gives detailed schemes for the preparation of standard test solutions.

1.7 Specifications and Reports

For the quality control of pharmaceutical products, assay design and technique must be of accuracy and precision appropriate to the purpose. This is considered in detail in Chapters 8 and 9.

A product specification is a prerequisite for meaningful quality control, for without defined acceptable lower (and sometimes upper) limits in terms of activity per unit weight (e.g., IU/mg) how can a product be assessed on the basis of determined potency?

It follows that the sample or a portion of it must be weighed before diluting accurately to prepare the test solution. This may seem to be an unnecessary statement of the obvious. It is a fact, however, that in some laboratories a decision to pass or reject a production batch is based on perhaps two assays in each of which the whole content of a unit dosage form is diluted for assay without weighing. Results are then expressed in terms of percentage of the amount of active substance declared on the label.

Disregarding for the moment assay errors in the estimated potency, how can the analyst distinguish between the cases (1) a low fill weight with material of pharmacopoeial quality, and (2) a correct fill weight with material of low potency?

Does a report of "100% of labeled claim" represent a correct fill of good quality material or an overfill of low potency (below pharmacopeial standards) material?

The former is satisfactory, the latter quite unsatisfactory. The analyst has no way of distinguishing between the two possibilities and so the assay has been done in vain.

Assay errors must now be considered in conjunction with weight variation of the unit dosage form. If the sample has not been weighed, the two sources of variation are indistinguishable. Widely discrepant results are often ascribed to "biological error" and accepted regardless, or the assay may be repeated in the hope of better luck next time!

A further complication arises in the case of formulations containing one or more active ingredients plus excipients. There is the possibility of poor blending of the ingredients, leading to unequal distribution between the individual doses.

The correct procedure of course in a manufacturer's quality control laboratory is to ensure that the bulk material is of satisfactory quality before labor and packaging materials are expended on the filling operation. There is normally no need to check potency of the filled material in routine production control. Correct dosage is controlled by rigorous checks on mean weight and weight variation of contents of vials.

With current medicines legislation in many countries requiring that product formulas be disclosed to a licensing and regulatory authority, there seems to be no good reason why laboratories of that authority should not avail themselves of this information and test according to a specification that defines potency without ambiguity in terms such as units of activity per milligram of the formulated product as packed by the manufacturer.

For meaningful quality control, whether by biological, chemical, or physical methods, it is highly desirable for the analyst

(1) to know the essential intended composition of the product,
(2) to have provided for him, or alternatively to devise for himself, a specification setting out quantitative limits appropriate to the intended composition,
(3) to express his results quantitatively in terms such as IU/mg, IU/ml, μg/mg, μg/ml, %w/w, or %w/v,
(4) when appropriate, to relate these results to weight or volume of individual doses, volume that can be withdrawn by a syringe, etc.

Kirshbaum (1972) explains how disagreement between a manufacturer and the regulatory authority (USFDA) may arise due to their different approaches to sample preparation. He attributes some disagreements to the manufacturer measuring, for example, the total antibiotic content of a vial, whereas the regulatory laboratory measures the amount of antibiotic that would be withdrawn by syringe in clinical use. It is the author's experience, however, that the majority of manufacturers are well aware of the need for overages. Such disagreements would be obviated by preparation of a realistic product control specification.

Expression of results by manufacturer and regulatory laboratory merely in terms of conformity with a label declaration seems to be a relic of the various food and drug laws of half a century ago. It is not an adequate criterion by which to judge the medicines of today.

References

Coffey, H. F., and Kuzel, N. R. (1966). U.S. Pat 3, 316, 854.
Gerke, J. R., Haney, T. A., and Pagano, J. F. (1962). *Ann. N.Y. Acad. Sci.* **93**, 640.
Grimshaw, J. J., and Jones, A. (1970). *Analyst* **95**, 466.
Grove, D. C., and Randall, W. A. (1955). "Assay Methods of Antibiotics, A Laboratory Manual," Medical Encyclopedia, New York.
Haney, T. A., Gerke, J. R., Madigan, M. E., Pangano, J. F., and Ferrari, A. (1962). *Ann. N.Y. Acad. Sci.* **93**, 627.
Jones, A., and Palmer, G. (1970). *Analyst* **95**, 463.
Jerne, N. K., and Wood, E. C. (1949). *Biometrics* **5**, 273–299.
Kavanagh, F. W. (1971). *J. Pharm. Sci.* **60**, 1858.
Kavanagh, F. W. (1974). *J. Pharm. Sci.* **63**, 1463.
Kirshbaum, A. (1972). *In* "Quality Control in the Pharmaceutical Industry" (M. S. Cooper, ed.), Vol. 1. Academic Press, New York.
Kuzel, N. R., and Kavanagh, F. W. (1971a). *J. Pharm. Sci.* **60**, 764.
Kuzel, N. R., and Kavanagh, F. W. (1971b). *J. Pharm. Sci.* **60**, 767.
Lightbown, J. W. (1961). *Analyst* **86**, 216.
Miles, A. A. (1952). *In* "Microbial Growth and its Inhibition," pp. 131–147. W. H. O., Monograph Series No. 10.
Shaw, W. H. C., and Duncombe, R. E. (1963). *Analyst* **88**, 694.
Stoddart, G. A. (1972). Personal communication.
Wright, W. W. (1971). *In Colloq. Int. Pharm. Res. St., Congr. Pharm. Sci., 31st, Washington D.C.*

CHAPTER 2
THE AGAR DIFFUSION ASSAY

2.1 Introduction

Inhibition zones in inoculated agar media have been used qualitatively for many decades to demonstrate antibacterial activity.

Quantitative measurements were made by Chain and his colleagues in 1940 to monitor purification processes in the isolation of penicillin.

This relationship between applied dose of antibiotic and the size of the resulting inhibition zone has remained for over thirty years as a basis for the comparison of samples of unknown potency with standard reference substances. The agar diffusion method, as it is called, is now used extensively in quality control laboratories throughout the world. It is a method potentially capable of yielding reliable potency estimates. This potential is achieved when assay design and practical techniques take into consideration the many factors other than dose of applied antibiotic that influence zone size.

All too often, unfortunately, the method is applied empirically using designs and practical techniques incapable of achieving the method's potential. Gross discrepancies between replicate determinations are ascribed to "biological error".

Kavanagh (1972a) has stated: "the antibiotic diffusion assay is not a biological assay. It is a physicochemical method in which a microorganism is used as an indicator. Most of the observed variations are caused by neglect of the physicochemical aspects and not by biological variation."

2.2 Theory of Zone Formation

The principles involved in the formation of inhibition zones in the antibiotic agar diffusion assay have been studied by several workers during the years 1946–1952.

Cooper and Woodman (1946), in a study designed purely to elucidate principles of diffusion under the conditions of this assay method, worked with crystal violet diffusing through agar gel in tubes. Mitchison and Spicer (1949) used narrow tubes (3 mm diameter) of inoculated medium for the routine assay of streptomycin. They concluded on both theoretical and practical grounds that the square of the width of the inhibition zone was approximately linearly related to the logarithm of the dose of streptomycin.

This technique was adopted by Cooper and Gillespie (1952) to study the influence of temperature on zone formation, and by Cooper and Linton (1952) to compare the results obtained in tubes with those from plates (see Fig. 2.1).

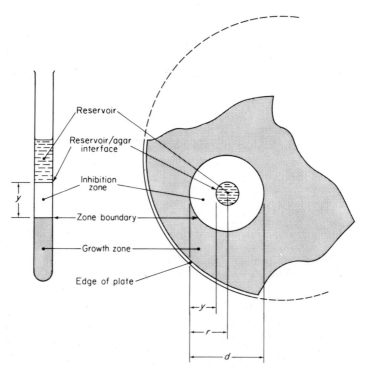

Fig. 2.1. Inhibition zones in tubes (left) of agar medium as used by Mitchison and Spicer (1949) and plates (right) of agar medium as used in most routine assay procedures. y, distance from reservoir/agar interface to zone boundary (tubes and plates); r, zone radius (plates); d, zone diameter (plates).

Work in this field is reviewed by Cooper (1963, 1972) in his contributions to Kavanagh's Volumes I and II of "Analytical Microbiology." These studies consider the mathematics of diffusion of the antibiotic through the agar gel, growth pattern of the test organism, and interaction of the antibiotic with the living cell.

The general system considered by various workers comprised:

(1) A reservoir from which the antibiotic was able to diffuse outward through the agar gel, thus inhibiting growth.

(2) A nutrient agar medium inoculated with a uniform suspension of a test organism either in the vegetative or spore form; this on incubation, after

a lag period or in the case of spores a germination plus lag period, multiplied to a level at which there was sufficient cell material to absorb all antibiotic, thus preventing further outward diffusion of antibiotic and so limiting the size of the inhibition zone.

Mathematical models are demonstrated for both linear diffusion and radial diffusion from a reservoir of constant concentration (m_0). The former is applicable when the reservoir is relatively large (8 mm diameter or more), the latter to small reservoirs, or beads or disks applied to the surface. In both cases equations are derived that show that there is a concentration gradient with continuously decreasing concentrations at increasing distances (y) from the reservoir/agar boundary.

When the reservoir is small, however, the concentration in the reservoir soon drops below its original value (m_0). Instead of a continuously decreasing concentration at increasing distances (y), the reservoir is surrounded by an expanding concentric peak concentration. This may lead to double inhibition zones.

Separate formulas representing diffusion (1) from a constant concentration and (2) from a falling concentration lead to the same conclusion. That is, the square of the parameter measured is directly proportional to the logarithm of m_0, for the parameters y in case (1) and the zone radius r in case (2). When reservoirs are small and zones large, r approximates to y.

Cooper has demonstrated certain mathematical concepts of importance for an understanding of the fundamental principles of this assay method. These concepts are very relevant to assay design and technique:

Critical concentration m':   the concentration of antibiotic arriving at the position of the future zone boundary at a certain time T_0 (see below).

Critical time T_0:   the period of growth of the organism at which it reaches the critical population N' (see below)

Inoculum population N_0:   the population at the time of inoculation.

Critical population N':   the population at time T_0 (further increase in population beyond this limit results in an excess of organism capable of completely absorbing the antibiotic and thus preventing its further outward diffusion; however, diffusion of the antibiotic during the lag phase of growth may result in small inhibition zones even if a very heavy inoculum were used such that $N_0 = N'$).

Inhibitory population N'':   that population which is just sufficiently large to completely prevent formation of inhibition zones.

The nature of these parameters may now be described in more detail.

Critical concentration m' is a measure of the sensitivity of the test organism under the particular assay conditions. It is not the same as minimum inhibitory concentration (m.i.c.), being about 2 to 4 times as great as m.i.c., which is determined under very different conditions. It is defined mathematically by

$$\ln m' = \ln m_0 - y^2/4DT_0 \qquad (2.1)$$

where D is the diffusion coefficient (expressed as mm/hour), which is dependent on temperature and viscosity of the medium and varies inversely as the radius of the molecule, and m_0 is the initial concentration of antibiotic. T_0, m' and y have the same meanings as before.

Critical concentration may be evaluated by plotting y^2 against $\log m_0$ (it is convenient to use logarithms to the base 10). The intercept of the straight line on the $\log m_0$ scale at $y^2 = 0$ corresponds to $\log m'$ (see Fig. 2.2). It is clear that concentrations below m' cannot produce inhibition zones.

Critical time T_0 is the time at which the position of the zone boundary is fixed. It may be determined by preincubation of the inoculated medium for

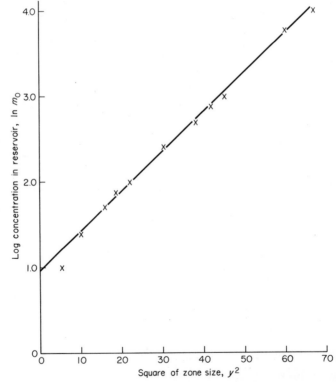

Fig. 2.2.   The relationship between the square of the width of inhibition zone y^2 and logarithm of applied dose of streptomycin ($\log m_0$) on agar medium in tubes inoculated with *Staphylococcus*. The critical concentration m' corresponds to the point of intersection of the response line with the log dose scale at zero response. Thus, critical concentration is obtained as antilog $0.95 = 8.9$ μg/ml. From Cooper (1963).

varying periods prior to addition of antibiotic solution to the reservoir. Critical time is defined mathematically by

$$T_0 = h + y^2/4D \ln(m_0/m') \qquad (2.2)$$

where h is the time of preincubation, and D, y, m_0, and m' have the same meanings as before. It is obtained by rearranging Eq. (2.1) and replacing T_0 by $T_0 - h$—a legitimate change since preincubation has effectively changed N_0. It may be evaluated by plotting y^2 against h for a fixed concentration m_0. The intercept of the straight line on the h scale at $y^2 = 0$ corresponds to $h = T_0$.

Cooper has shown that repetition at different concentrations gives the same value for T_0 (see Fig. 2.3).

Critical time is dependent on the lag period L and the generation time G. It is therefore temperature dependent. It is also dependent on the inoculum level N_0. It is independent of antibiotic concentration.

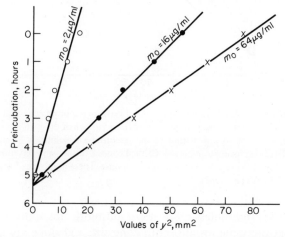

Fig. 2.3.   Graphs showing the determination of critical time T_0. The three plots of square of zone width (y^2) (corresponding to three concentrations of antibiotic) versus period of pre-incubation all intersect at the same point on the time scale (5.4 hours) corresponding to $y^2 = 0$. This demonstrates that the critical time is independent of the dose of antibiotic and that for these experimental conditions (test organism, inoculum level, medium, and incubation temperature) it is 5.4 hours. The data are from the work of Cooper (1963) using streptomycin and *Staphylococcus.*

To summarize, the factors ultimately deciding the position of the zone boundary are critical concentration and critical population. Factors that influence the diffusion of the antibiotic or the time of achieving the critical population therefore affect the zone size.

It should be noted that in accordance with the theory of inhibition zone formation the slope of the response line is determined by the following factors: (1) preincubation (see Fig. 2.4), (2) prediffusion, and (3) temperature of incubation (diffusion coefficient increases with temperature, whereas growth rate increases to an optimum and then falls).

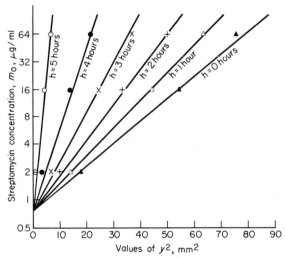

Fig. 2.4.   The relationship between logarithm of dose (of streptomycin), and square of zone width (y^2) in agar medium inoculated with *Staphylococcus*, when the medium is preincubated for varying periods prior to application of the streptomycin solutions. Each response line represents a different preincubation period. Critical concentration (m') is given by the common point of intersection of the response lines at zero response. These graphs are from the same data as used for Fig. 2.3 (Cooper, 1963).

It is clear then that unless all doses of antibiotic are applied to the inoculated agar at the same instant, both zone size and slope are liable to unwanted influences. Practical techniques must take this into account.

The influence of inoculum level is shown graphically in Figs. 2.5a and 2.5b. It is seen that lower inoculum level leads to larger zone sizes and better slopes of the log dose–response lines, i.e., greater contrast in responses to a fixed dose interval. However, if inoculum level is reduced too far, then zone boundaries become diffuse. Choice of inoculum level must therefore be a compromise designed to give sharply defined zone boundaries with a good slope of the log dose–response line.

The interaction of these two assay characteristics is discussed in Chapter 8.

Cooper's Eq. (2.1), being based on a study of linear diffusion in tubes of agar, does not include any term that can explain the influence of thickness

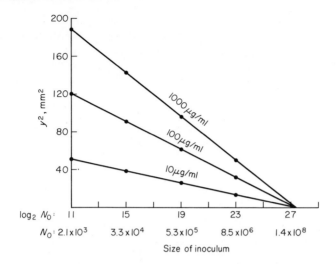

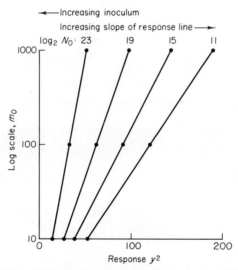

Fig. 2.5. (a) The effect of increasing inoculum (*Staphylococcus*) on square of zone width (y^2) is shown for three dose levels of streptomycin. The three dose levels converge at zero response at an inoculum level corresponding to N''. (b) The same data are replotted as dose–response lines (y^2 versus logarithm of dose). The decreasing slope of the dose–response lines with increasing inoculum level is of great practical significance. (From Cooper, 1963.)

of the agar layer in a plate assay. Humphrey and Lightbown (1952) showed
the influence of agar thickness by a study of diffusion from small beads placed
on the surface of the agar medium in a dish. They derived the equation

$$r^2 = 4DT[\ln M/H - \ln C' - \ln(4\pi DT)] \qquad (2.3)$$

where *r* is the radius of zone of inhibition, *H* the thickness of the agar layer,
M the quantity of antibiotic, *C'* the concentration of antibiotic at the visible
edge of the zone, *T* the time of diffusion, and *D* the diffusion constant.

It is seen from the term $\ln M/H$ that increased thickness of the agar will
lead to smaller zones.

Lees and Tootill (1955a) have summarized the factors affecting zone size
in both small and large plate assays as follows:

(1) *Choice of test organism*:   Its inherent sensitivity.

(2) *Condition of test organism, vegetative or spore*:   If vegetative, its
phase of growth.

(3) *Density of seeding*:   The zone width is related inversely to the size
of inoculum (see Figs. 2.5 and 2.6)

(4) *Formulation and condition of the medium*:   A rich medium results
in more rapid growth with consequent smaller zone sizes; water content of
the medium: drying out of the agar gel at the edges of large plates may lead
to inflated zone diameters.

(5) *Thickness of the agar medium*:   As thickness increases zone width
decreases.

(6) *Potency of the test solution.*

(7) *Volume of the test solution applied to the plate*:   This should be large
enough to act as a reservoir of constant concentration or should be a standard volume.

(8) *Area of seeded agar to which the test solution is applied*:   A larger
stainless steel cylinder, bead, or paper disk on the surface of the agar or
cylindrical well cut out of the agar naturally leads to a zone of greater
diameter.

(9) *Time of application of the test solutions*:   Solutions applied to the
plate appreciably later than the first have less time for diffusion before the
critical population is achieved, thus producing relatively smaller zones.

(10) *Temperature of incubation of assay plate*:   More rapid growth results
in smaller zone sizes so that uniformity of incubation temperature is of great
importance.

Of these factors, those liable to contribute to varying responses to a single
test solution are (3)–(5) and (7)–(10). Practical steps may be taken to minimize
these influences. Thus, flat-bottomed dishes should be selected. These dishes

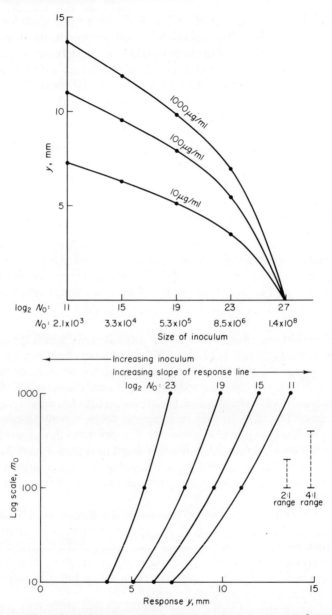

Fig. 2.6. (a) The data of Fig. 2.5a replotted, this time using y in place of y^2. (b) The same data plotted in the form of dose–response lines, as y versus logarithm of dose. This is the form commonly used in routine assays. It is seen that curvature appears to be slight over a short dose range such as 2:1 or 4:1. As in the case of Fig. 2.5b, the decreasing slope with increasing inoculum is apparent.

should be placed on a level surface (checked with a spirit level) for pouring in the molten media. The inoculum should be uniformly dispersed by thorough mixing. Volume of medium should be accurately measured. Timing of the addition of different test solutions should be so regulated as to minimize differences in time or alternatively to compensate for them. Test solutions should all be at the same temperature. Incubation conditions should be such as to ensure uniform heating rates for all plates and all parts of individual plates.

Operations designed to improve accuracy are described by Kavanagh (1972b, 1974). See also Section 2.10, in which large plates are described.

2.3　Nature of the Response Curve

The studies reviewed by Cooper have contributed greatly to an understanding of the principles of the diffusion assay. The concepts of critical concentration and critical time are based on a rectilinear relationship between logarithm of applied dose and square of y, the width of the inhibition zone.

Bryant (1968) describes routine assay procedures in which tubes of agar medium are used and square of zone width is the basis of potency calculation. However, by far the greater part of routine potency determinations are by the plate method, and the observed response is zone diameter d. Calculation procedures normally assume a direct relationship between d and logarithm of applied dose. This may at first sight appear to be a contradiction of the theories previously expounded. However, Cooper notes deviations from the y^2 versus log dose relationship when y is small (less than 3 mm). Mitchison and Spicer (1949) derive equations showing that when y is large, y^2 is proportional to log dose but that when y is small then y itself is proportional to log dose.

Humphrey and Lightbown (1952) considered diffusion from beads on the agar surface and concluded that d^2, the square of the zone diameter, should be proportional to log dose. A plot of zone diameter itself against log dose should therefore theoretically be curved. However, they state that such curvature would be slight and only detectable if dose intervals were large or if a very accurate assay were performed.

Reconsidering the work of Cooper and Woodman, their data of Fig. 2.5a may be replotted in the form y^2 versus log dose. It is seen (Fig. 2.5b) that this results in a series of straight dose–response lines and that with increasing inoculum both zone size and slope of the dose–response line are reduced.

As in practice the normal relationship with log dose is not y^2 but (effectively)

y. It is then necessary to consider Figs. 2.6a and 2.6b, in which the data of Figs. 2.5a and 2.5b are replotted using y in place of y^2.

It is seen from Fig. 2.6b that although a plot of y versus log dose is curved, if the overall dose range is only 2:1 or 4:1, it appears to approximate a straight line.

Experience with many antibiotic agar diffusion assays confirms that this approximation forms a reasonable basis for calculation of potencies. In practice, overall dose ratios are often 2:1 or 4:1. However, in Examples 1 and 7 the approximation still seems valid even though the overall ratio is 8:1. It is shown in Chapter 6 that the internal evidence of the assay confirms the validity of the assumption.

In Example 6 (which is a vitamin assay), an overall dose ratio of 10:1 is used. As this is only a two dose level assay, no evidence of validity is possible from the data of the assay. However, previous experience of the method had indicated a linear relationship between log dose and zone diameter over a range wider even than this.

As pointed out by Humphrey and Lightbown, detection of curvature is also dependent on the precision of the observations. If random errors are relatively large, curvature, even though a fact, may be indistinguishable from random error.

Statistical tests for curvature are demonstrated in Chapter 6.

The author has on occasion observed curvature in a plot of zone diameters (d) versus log dose in three dose level assays with an overall dose range of 4:1. In such cases d^2 versus log dose often gave a linear plot. It was found, however, that whether d or d^2 were used for the potency calculation the result was virtually the same, differences being only about 0.1%.

In Chapter 8 the influence of quadratic curvature is discussed. It is shown that when balanced assay designs are used, such curvature is entirely without influence on the calculated potency.

It seems therefore that use of the relationship zone diameter versus log dose is very often a valid basis for the calculation of potency ratios.

2.4 Dose–Response Curves in Practice

Observations typical of those obtained in an agar diffusion assay for an antibiotic are illustrated by Example 1. In this test, six petri dishes were used and four inhibition zones corresponding to the four different concentrations of dose of a standard preparation (streptomycin) were developed on each dish. The observed responses (d), i.e., zone diameters in millimeters, are tabulated.

The purpose of this exercise was solely to investigate the nature of the dose–response curve so as to select a convenient range for future work. No samples were assayed using this standard curve.

Replication of doses to balance out random errors in responses and obtain a mean more reliable than any individual observation is essential in the agar diffusion assay, where the error of a single observation may be large. For much routine work a replication of six is convenient.

Example 1: A Standard Curve for Streptomycin

The test organism was *Bacillus subtilis* ATCC 6633. Test solutions were applied to the single-layer medium by the ceramic bead method. Observed zone diameters (in millimeters) are given in Table 2.1. The mean responses from Table 2.1 are plotted against dose in Fig. 2.7a and against logarithm of dose in Fig. 2.7b. In the latter, the four points approximate to a straight line. It is assumed that the graph (Fig. 2.7b) represents the best straight line that can be fitted to these four points. Its slope can then be expressed mathematically by using the points read from the graph that correspond to low and high dose, respectively. Thus,

$$\text{mean response to low dose} = 13.33 \text{ mm}$$

$$\text{mean response to high dose} = 19.30 \text{ mm}$$

These extreme doses are separated by three intervals of log 2, i.e., 0.301; therefore, an increase in log dose of 0.301 corresponds to a mean increase in response of $(19.30 - 13.33)/3 = 1.99$ mm.

It will be seen later that for statistical evaluation it is more convenient to express the slope as b, a hypothetical response difference over a tenfold increase in dose. In this case,

$$b = (1.99 \times \log 10)/\log 2 = 1.99 \times 1/0.301 = 6.61 \text{ mm}.$$

Table 2.1

Dose (IU/ml):	80	40	20	10
	19.3	17.2	15.3	13.2
	19.9	17.8	15.8	13.5
	19.6	16.6	14.5	13.5
	18.7	16.7	14.7	12.9
	19.3	17.3	15.2	13.1
	19.6	17.9	16.2	13.8
Total:	116.4	103.5	91.7	80.0
Mean:	19.40	17.25	15.28	13.33

The value 6.61 mm has been obtained from a graph judged by eye to be the best straight line fitting the four points. As will be shown in Section 2.9,

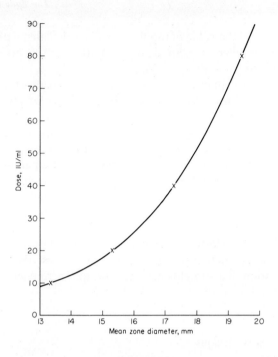

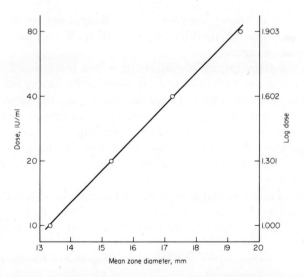

Fig. 2.7. (a) A typical relationship between dose of antibiotic and response (diameter of inhibition zone). The data are from Table 2.1. (b) Mean response (zone diameter) is plotted against log dose.

the best estimate of the value of the slope can be calculated without drawing the graph. Three different methods of calculation are illustrated and these all give values of 6.70 mm.

It should be noted that in the United States of America the term "slope" is generally taken to mean the increase in zone diameter for a twofold increase in dose.

The disadvantages of working from a curvilinear response line such as that shown in Fig. 2.7a are clear. It is difficult to judge what is the best curve to fit the observations. In this example, curvature is not apparent in the plot of zone diameter against log dose, and so this is a very much more convenient form in which to work with the observations. Fortunately, this same transformation leading to an approximately straight line is applicable to the assays of many antibiotics and growth-promoting substances. It is of course necessary to check the form of the log dose–response line in each case before using any of the calculation procedures that will be described and are based on the assumption of a straight-line relationship. The influence of curvature on such calculations is discussed in Section 8.4.

2.5   Simple Assay Designs

Potencies of unknown samples may be determined by measuring the response to an appropriate dilution and reading off a standard curve such as described in Example 1.

Due to the many extraneous factors described in Section 2.3 that affect response, day to day variations make it essential that each assay be a direct comparison of sample with reference standard under the same conditions.

While it is perfectly legitimate to compare sample responses with a standard response line determined at the same time, it is often preferable to use an alternative design of experiment that will provide: (1) better results for the same amount of practical effort, and (2) some indication of the validity of the assay.

Many designs are available from which one may be selected, taking into consideration: (1) the nature of the samples, (2) the number of the samples, (3) the required reliability of the result. Choice of design is discussed in Chapter 8.

The special characteristics of the log dose–response rectilinear relationship are:

(1)   The response line of any other qualitatively identical substance should be parallel to that of the standard, from which it follows that the slope may be measured not only from the standard but also from the sample.

(2)   It is essential to have at least two dose levels to estimate the slope of the response line. (Contrast with linear arithmetic responses, where one observation and the zero origin might suffice.)

(3)   In the event of the sample and standard response lines not being parallel (discounting minor deviations due to experimental error) potencies of test solutions would appear to be identical at the point of intersection. Such an assay would be invalid, with estimated sample potency greater than standard above the intersection and lower than standard below the intersection, or vice versa.

Knudsen and Randall (1945) described a simple and efficient design for the assay of penicillin. This design employs two dose levels each for standard and sample. The ratio between high and low doses is the same for both these preparations. It is known as a 2 + 2 design. Each petri dish includes one of the four doses (treatments), and so the replication is equal to the number of dishes and is the same for all treatments. Such a design is said to be "symmetrical" or "balanced."

This simple 2 + 2 is included in the International Pharmacopeia. It and its various modifications for the simultaneous comparison of two or more samples with the same standard are widely used today.

The principle of the calculation procedure may be seen by reference to Fig. 2.8, which is a graphical representation of the dose–response relationship for two preparations, standard and sample. Test solutions for the two preparations have the same *nominal* potency. However, their difference in actual (log) potency is revealed by the vertical distance between the two parallel lines.

Mean responses (zone diameters) are calculated for the corresponding treatments on all plates thus: S_1 = mean response to standard low dose, S_2 = mean response to standard high dose, T_1 = mean response to sample low dose, T_2 = mean response to sample high dose.

Assuming that the lines are essentially parallel but mean responses are subject to random error, then the best estimate of the difference in response due to difference between high and low doses is obtained as the mean of these differences for standard and sample and is designated E:

$$E = \tfrac{1}{2}[(S_2 + T_2) - (S_1 + T_1)]$$

Similarly the best estimate of the difference in response due to difference between sample and standard is obtained by averaging the differences at the two levels and is designated F:

$$F = \tfrac{1}{2}[(T_1 + T_2) - (S_1 + S_2)]$$

From the graph (Fig. 2.8) it is seen that the difference in log dose corresponding to E is I, the log dose ratio, and that corresponding to F is M, the

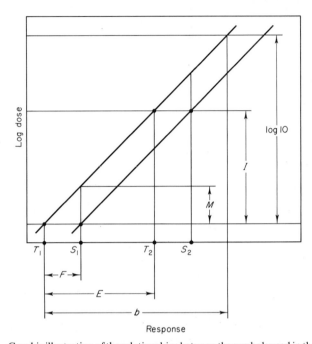

Fig. 2.8. Graphic illustration of the relationships between the symbols used in the calculation of the potency ratio (unknown versus standard) in parallel line assays. S_1 = mean response to low dose of standard, S_2 = mean response to high dose of standard, T_1 = mean response to low dose of unknown, T_2 = mean response to high dose of unknown, E = mean difference in response due to difference between adjacent dose levels, F = mean difference in response due to difference between unknown and standard doses, b = calculated hypothetical increase in response to a tenfold increase in dose, I = logarithm (base 10) of the ratio of adjacent dose levels, M = logarithm (base 10) of the ratio of unknown to standard potency.

logarithm of the ratio of actual sample potency to standard potency. The graph also shows the symbol b, which is the calculated difference in response (assuming a straight line) that corresponds to a dose interval of ratio 10:1.

From a consideration of similar triangles it is readily seen that

$$F/M = E/I = b/\log 10 = b \qquad (2.4)$$

from which

$$M = F/b \qquad (2.5a)$$

In this case where sample and standard are of the same *nominal* potency, then M also corresponds to the difference in *actual* log potency of sample from its *nominal* log potency. The corresponding potency ratio is obtained simply by taking the antilog of M.

It should be noted that the principles described here for the simplest case, the 2 + 2 design, are applicable to all parallel line assays. When nominal doses are unequal, however, an additional step must be introduced.

Suppose that sample and standard are at differing nominal potencies, i.e., $\bar{x}_T \neq \bar{x}_S$. In this case the difference in logarithms of sample actual to sample nominal dose M' is obtained via M by

$$M' = M - (\bar{x}_T - \bar{x}_S) = M + \bar{x}_S - \bar{x}_T \qquad (2.5b)$$

The calculation for such cases is illustrated by Examples 7, 13, and 17.

Returning now to the simple case of the 2 + 2 design, the convenience of this arrangement is illustrated by Example 2, the assay of streptomycin using *Bacillus subtilis* as test organism. The actual dose–response relationship for this assay is shown in Fig. 2.9.

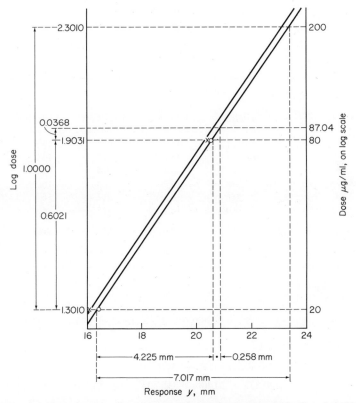

Fig. 2.9. Log dose–response lines from the assay of streptomycin by 2 + 2 design using *Bacillus subtilis* (Example 2). ×, mean responses to standard doses; ○, mean responses to sample doses. The lines are extrapolated so that all parameters of the calculation may be illustrated. $E = 4.225$ mm, $F = 0.258$ mm, $b = 7.017$ mm, $I = 0.6021$, $M = 0.0368$.

Example 2:   Assay of Streptomycin

Test organism: *Bacillus subtilis* ATCC 6633
Dose ratio: 4:1
Standard: streptomycin sulfate, potency 745 IU/mg
Weighings and dilutions to high-dose test solution (80 IU/ml):

$$70.2 \text{ mg } (52{,}299 \text{ IU}) \rightarrow 65.4 \text{ ml}: 10 \text{ ml} \rightarrow 100 \text{ ml}$$

Sample: streptomycin sulfate injection containing nominally 1 g streptomycin base per vial

Weighing and dilutions to high-dose test solution: total vial contents 1.4692 g,

$$1.4692 \text{ g} \rightarrow 500 \text{ ml} : 10 \text{ ml} \rightarrow 250 \text{ ml}$$

Nominal test solution potency: 80 IU/ml

From both standard and sample high dose test solutions, 1:4 dilutions were prepared for the low dose. Test solutions were applied by the ceramic bead technique. The response (zone diameters in millimeters) is given in Table 2.2.

Table 2.2

	Unknown sample		Standard	
	High dose	Low dose	High dose	Low dose
	20.1	15.7	19.8	15.3
	20.9	16.5	20.7	15.9
	20.9	16.4	20.4	16.6
	20.8	16.7	21.0	16.3
	20.6	16.8	20.2	16.4
	19.9	16.5	20.3	15.8
Treatment total	123.2	98.6	122.4	96.3
Treatment mean	20.533	16.433	20.400	16.050

Thus, from Table 2.2

$$S_1 = 16.050, \qquad S_2 = 20.400, \qquad T_1 = 16.433, \qquad T_2 = 20.533$$

Difference due to dose:

$$E = \tfrac{1}{2}[(T_2 + S_2) - (T_1 + S_1)]$$
$$= \tfrac{1}{2}[(20.533 + 20.400) - (16.433 + 16.050)] = 4.225$$

Difference due to sample:

$$F = \tfrac{1}{2}[(T_2 + T_1) - (S_2 + S_1)]$$
$$= \tfrac{1}{2}[(20.533 + 16.433) - (20.400 + 16.050)] = 0.258$$

log ratio of doses:

$$I = \log 4 = 0.6021$$

Slope:

$$b = E/I = 4.225/0.6021 = 7.0171$$
$$M = F/b = 0.258/7.0171 = 0.0368$$

Potency ratio:

$$\text{antilog } M = \text{antilog } 0.0368 = 1.088$$

Thus the potency of the high dose test solution is $1.088 \times 80 = 87.04$ IU/ml, leading to a sample potency of

$$(87.04 \times 250 \times 500)/(10 \times 1.4692 \times 1000) = 741 \text{ IU/mg}$$

Although not an essential part of the calculation, it is good practice to compare the values $S_2 - S_1$ and $T_2 - T_1$. In this case,

$$S_2 - S_1 = 20.400 - 16.050 = 4.350 \text{ mm}$$
$$T_2 - T_1 = 20.533 - 16.433 = 4.100 \text{ mm}$$

Ideally these two measures of slope should be identical. However, discrepancies such as this may be ascribed to random error. In contrast, a discrepancy of about 1 mm would suggest something basically wrong with the assay. Significant differences in slope might be caused by the active ingredient of the sample not being the same substance as the standard. Alternatively other ingredients in the sample may modify the response.

This assay is open to some criticism:

(1)   The standard is diluted to 65.4 ml so as to obtain a high-dose test solution of exactly 80 IU/ml, a quite unnecessary procedure, which is unfortunately a very common practice in microbiological assay. The awkward dilution, which is an additional source of error, could have been avoided by weighing accurately between 50 and 60 mg of standard, diluting to 50 ml and applying a factor as is standard practice in volumetric analysis, e.g.,

$$58.3 \text{ mg} \rightarrow 50 \text{ ml} : 10 \text{ ml} \rightarrow 100 \text{ ml}$$

giving an actual high-dose test solution potency of

$$(58.3 \times 745 \times 10)/(50 \times 100) = 86.87 \text{ IU/ml}$$

which can be described as 80 IU/ml ($f = 1.086$).

(2)   Weighing such a large amount of sample necessitates large dilutions, which are extravagant in the use of buffer solution. It would be more convenient to weigh about 110 mg of sample and dilute, thus:

$$110 \text{ mg} \rightarrow 100 \text{ ml} : 10 \text{ ml} \rightarrow 100 \text{ ml}$$

The custom of weighing the whole contents seems to be related to the intention of comparing the total number of units in the vial with the labeled claim. Certainly, the weight of contents should be determined. For assay purposes, however, a smaller weight is more convenient.

(3)   The dose levels are rather high. While this has the advantage that samples require less dilution, it might result in interference in the case of samples containing a high proportion of excipients.

The European and International Pharmacopeias give analogous formulas for other designs of assay. For these other assays the principles of the calculations are identical, although in some cases the underlying logic is a little more complicated.

For routine work it is usually simpler to work with total zone diameters rather than averages.

This is a perfectly legitimate step yielding the same result for less labor. However, for the statistical analysis that will be described later, the value of the slope b must be calculated on the basis of mean zone sizes, i.e., $b =$ calculated mean difference in zone diameter between responses due to doses having a ratio of 10:1.

2.6   Simple Multiple Assay Designs

Laboratories examining a large number of similar samples may often find it convenient to use a design incorporating two or more samples with each standard. This reduces the amount of practical effort and as each sample that is used at more than one dose level contributes to the determination of slope, a better estimate of slope is obtained.

In fact, the observations quoted in Example 2 were taken from an assay including two samples and one standard each at two dose levels. Thus in each petri dish there were six zones corresponding to the six treatments. These were distributed in accordance with randomized designs such as are shown in Appendix 1.

In Example 3 the calculation is modified to include the contribution of the additional sample to the estimate of slope.

Example 3:   Assay of Streptomycin

Test organism: *Bacillis subtilis* ATCC 6633
Dose ratio: 4:1
Standard and sample 1:   dilutions are identical with those of Example 2
Sample 2:   a solution containing nominally 10 mg streptomycin base/ml, dilution to high-dose test solution:

$$10 \text{ ml sample} \rightarrow 100 \text{ ml} : 20 \text{ ml} \rightarrow 250 \text{ ml}$$

From standard and sample high-dose test solutions (80 IU/ml), 1:4 dilutions were prepared for the low doses. Test solutions were applied to the single-layer medium by the ceramic bead method.

Responses are shown in Table 2.3

Table 2.3

Responses (zone diameter in mm) in the assay of streptomycin (Example 3)[a]

	Sample 1		Sample 2		Standard	
	1	2	3	4	5	6
	20.1	15.7	20.3	15.9	19.8	15.3
	20.9	16.5	20.5	15.8	20.7	15.9
	20.9	16.4	20.5	15.9	20.4	16.6
	20.8	16.7	20.2	16.2	21.0	16.3
	20.6	16.8	20.5	15.8	20.2	16.4
	19.9	16.5	20.1	15.7	20.3	15.8
Treatment total	123.2	98.6	122.1	95.3	122.4	96.3
Treatment mean	20.533	16.433	20.350	15.883	20.400	16.050

[a] Test solutions are numbered 1 to 6; odd numbers are high doses, even numbers are low doses.

Using the symbols as before and distinguishing mean responses, etc., to the second sample by T_2', T_1', F', and M':

$$S_1 = 16.050, \quad T_1 = 16.433, \quad T_1' = 15.833$$
$$S_2 = 20.400, \quad T_2 = 20.533, \quad T_2' = 20.350$$

Difference due to dose:

$$E = \tfrac{1}{3}[(T_2 + T_2' + S_2) - (T_1 + T_1' + S_1)]$$
$$= \tfrac{1}{3}(61.283 - 48.366) = 12.917/3 = 4.30567$$

Difference due to sample 1:

$$F = \tfrac{1}{2}[(T_2 + T_1) - (S_2 + S_1)]$$
$$= \tfrac{1}{2}(36.966 - 36.450) = 0.516/2 = 0.258$$

Difference due to sample 2:

$$F' = \tfrac{1}{2}[(T_2' + T_1') - (S_2 + S_1)]$$
$$= \tfrac{1}{2}[36.233 - 36.450] = -0.217/2 = -0.1085$$

Log ratio of doses:

$$I = \log 4 = 0.6021$$

Slope:

$$b = E/I = 4.3057/0.6021 = 7.1511$$

Sample 1:

Logarithm of potency ratio of T/S:

$$M = F/b = 0.258/7.1511 = 0.0361$$

Potency ratio of T/S:

$$\text{antilog } M = \text{antilog } 0.0361 = 1.087$$

corresponding to a sample potency of 740 IU/mg (compare these with results obtained in Example 2).

Sample 2:

Logarithm of potency ratio of T'/S:

$$M' = F'/b = -0.1085/7.1511 = -0.0152 = \bar{1}.9848$$

Potency ratio of T'/S:

$$\text{antilog } M' = \text{antilog } \bar{1}.9848 = 0.965$$

Sample potency:

$$(80 \times 0.965 \times 250 \times 100 \times 1)/(20 \times 10 \times 1000) \text{ mg/ml} = 9.65 \text{ mg/ml}$$

On comparing Example 3 with Example 2 it is seen that the use of the second sample in the calculation has made only a slight difference to the estimate of the slope b. It has had only a negligible effect on the estimated potency of sample 1. However, as will be seen in the section dealing with statistical evaluation (Chapter 6), it has caused a slight improvement in the precision and has therefore narrowed the confidence limits of the potency estimate.

It should be noted that in the calculation of E the fraction $\frac{1}{3}$ replaces the $\frac{1}{2}$ of Example 2. This is because three differences are being averaged and not two.

2.7 Simplification of Computation of the Potency Ratio

The labor in computation can be reduced by the use of tables relating the ratio F/E to R, the potency ratio. A separate table is required for each dose ratio. In Appendix 3, tables are given for both 2:1 and 4:1 dose ratios. The tables are quite simply constructed as will be exemplified by calculation of the F/E ratio for $R = 0.95$ in a 2:1 dose ratio assay.

From Fig. 2.8 it is seen by consideration of similar triangles that

$$F/E = M/I$$

Substituting $I = \log 2 = 0.301$ and $M = \log 0.95 = \bar{1}.9777 = -0.0223$,

$$F/E = -0.0223/0.301 = -0.074$$

Thus in a 2:1 dose ratio assay, when $F/E = -0.074$, $R = 0.95$. The method of use is shown by application to Example 3. The procedure is as follows: Obtain the ratio F/E, in this case for sample 1,

$$F/E = 0.258/4.306 = 0.060$$

Using the 4:1 dose ratio table, find the figure for F/E nearest to 0.060. Then the corresponding value of R gives the potency ratio. The table gives

F/E	R
0.056	1.08
0.062	1.09

The potency ratio may therefore be reported as 1.09. Interpolation gives the value 1.087, which agrees closely with the figure found by calculation. For sample 2,

$$F/E = -0.1085/4.306 = -0.025$$

The table gives

F/E	R
-0.029	0.96
-0.022	0.97

The potency ratio may be reported as 0.97. Interpolation gives the value 0.966, in good agreement with the calculated figure.

Proformas for recording observations and calculations are useful labor-saving aids. An example is given in Appendix 4.

2.8 Designs Incorporating Checks for Curvature (Three Dose Levels)

When preliminary work has shown that, under the standardized conditions used for the assay of any substance, a straight line log dose–response graph of adequate slope is obtained, it can be assumed that a similar (but not identical) line will be obtained in future routine work. That is, of course, dependent on assay conditions remaining substantially unchanged and samples being qualitatively similar.

However, in the preliminary work to establish an assay, it is customary to use an assay design that permits checking of standard and sample response lines for curvature. A design intended to check for curvature of the response line must obviously have three or more dose levels. The simplest of these designs is the 3 + 3 assay incorporating three dose levels of each standard and sample. For computational efficiency, dose levels must be in geometrical progression so that logarithms of dose form an arithmetical progression. For example, dose ratios of 1 : 2 : 4 or 4 : 6 : 9 would be suitable.

When a sample contains substances other than the ingredient to be estimated, it is possible that these extraneous substances might modify the response. Examination at more than two dose levels improves the chances of detecting such interferences, which might cause differences in slope or in curvature of the response lines for sample and standard. Such "drift" would be evidence of invalidity. In contrast however, absence of drift is not proof of validity.

Three dose level assays are commonly used when calibrating reference preparations. The necessity for this is questioned in Chapter 8, however.

Such a design for a three dose level assay is illustrated in Example 4, the assay of benzylpenicillin using *Sarcina lutea* as test organism. This assay was one of many in a collaborative program to establish a working standard. In such a series of tests it is desirable that the two preparations, master standard, and proposed working standard should be compared using more than one test organism. Significant differences in estimated potency with the different organisms would suggest a qualitative difference in composition of the two preparations, which might lead to unacceptability of the test preparation for use as a working standard.

For this reason *Sarcina lutea* assays were included in the collaborative exercise even though it was recognized that zone boundaries are not well defined in this assay. It should not be inferred that this is a test organism of choice for routine work.

Example 4 :   Assay of Penicillin

Test organism: *Sarcina lutea* ATCC 9341
Dose ratio: 2 : 1
Weighings and dilutions for high dose test solutions: International standard, sodium penicillin, 1670 IU/mg,

$$12.45 \text{ mg} \rightarrow 100 \text{ ml} : 5 \text{ ml} \rightarrow 100 \text{ ml} : 10 \text{ ml} \rightarrow 100 \text{ ml}$$

giving a standard solution of potency 1.040 IU/ml (nominal potency, 1 IU/ml).

Sample, potassium penicillin of nominal potency 1595 IU/mg,

$$26.07 \text{ mg} \rightarrow 100 \text{ ml} : 5 \text{ ml} \rightarrow 100 \text{ ml} : 5 \text{ ml} \rightarrow 100 \text{ ml}$$

Both sample and standard high dose test solutions were diluted further to give medium and low dose test solutions:

$$25 \text{ ml} \rightarrow 50 \text{ ml} \qquad (0.5 \text{ IU/ml})$$
$$25 \text{ ml} \rightarrow 100 \text{ ml} \qquad (0.25 \text{ IU/ml})$$

Test solutions were applied to the plate by means of the ceramic bead technique. Responses are shown in Table 2.4.

Table 2.4

Responses (zone diameter in mm) in the assay of penicillin (Example 4)[a]

	Sample			Standard		
Dose level:	High	Medium	Low	High	Medium	Low
Test solution number:	1	2	3	4	5	6
	24.4	20.7	17.4	24.0	20.4	17.0
	22.2	19.3	14.9	22.7	19.7	14.9
	22.3	18.0	15.0	22.0	18.6	15.0
	22.2	19.0	14.8	22.4	18.3	14.6
	22.6	17.8	14.4	22.3	18.0	14.7
	23.0	19.3	14.5	23.3	19.1	14.4
	22.4	19.4	15.0	22.5	19.0	14.9
Treatment total:	159.1	133.5	106.0	159.2	133.1	105.5
Treatment mean:	22.729	19.071	15.143	22.743	19.014	15.071

[a] Test solutions for high, medium, and low doses have nominal potencies of 1.00, 0.50, and 0.25 IU/ml, respectively.

Thus,

$$S_1 = 15.071, \qquad S_2 = 19.014, \qquad S_3 = 22.743$$
$$T_1 = 15.143, \qquad T_2 = 19.071, \qquad T_3 = 22.729$$

Difference due to dose:

$$E = \tfrac{1}{4}[(T_3 + S_3) - (T_1 + S_1)]$$
$$= \tfrac{1}{4}[(22.729 + 22.743) - (15.143 + 15.071)] = 3.8145 \text{ mm}$$

Differences due to preparations:

$$F = \tfrac{1}{3}[(T_3 + T_2 + T_1) - (S_3 + S_2 + S_1)]$$
$$= \tfrac{1}{3}[(22.729 + 19.071 + 15.143) - (22.743 + 19.014 + 15.071)]$$
$$= +0.0383 \text{ mm}$$

Log ratio of doses:

$$I = \log 2 = 0.3010$$

Slope:

$$b = E/I = 3.8145/0.3010 = 12.6728 \text{ mm}$$

Log of potency ratio (of T/S):

$$M = F/b = 0.0383/12.6728 = 0.00302$$

Potency ratio:

$$\text{antilog } M = \text{antilog } 0.00302 = 1.007$$

Sample potency:

$$(1.040 \times 1.007 \times 100 \times 100 \times 100)/(5 \times 5 \times 26.07) = 1607 \text{ IU/mg}$$

In the calculation of E, $\tfrac{1}{4}$ is used outside the bracket because the sum inside the bracket corresponds to the difference in response over two dose levels for two pairs of test solutions, i.e., four dose intervals in all. Thus E is obtained as the mean of four differences. Similarly in the calculation of F, $\tfrac{1}{3}$ is used outside the bracket because the sum inside the bracket is that of three differences between sample and standard.

Note that in the calculation of E, difference due to dose, the medium dose mean responses T_2 and S_2 are not used. The reason is readily seen by expressing the slope as that due to difference between high and medium dose plus that due to difference between medium and low dose. Thus for the standard, slope corresponds to

$$(S_3 - S_2) + (S_2 - S_1) = (S_3 - S_1)$$

Observations at the exact midpoint of the dose levels make no contribution to the estimate of slope. The usefulness of these midpoint observations is (1) to show curvature if it is present, and (2) to contribute to the comparison of sample and standard.

The results of this assay are expressed graphically in Fig. 2.10. It will be seen that the two lines deviate very slightly from parallelism, also that the observations corresponding to medium dose lie slightly to the right of the response lines joining high and low responses. These slight deviations from the ideal should not be interpreted as signifying invalidity of the assay.

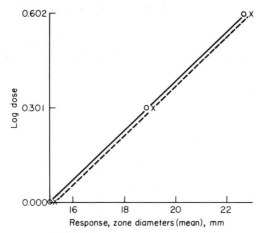

Fig. 2.10.   Log dose–response lines from the assay of penicillin by 3 + 3 design using *Sarcina lutea* ATCC 9341 (Example 4); (○) mean response to standard doses, (×) mean response to sample doses.

In this case, where the deviation from parallelism and linearity are so slight, it seems obvious that they can be attributed to random experimental error. A statistical evaluation (Chapter 6) confirms that this is a reasonable assumption.

2.9   Designs Incorporating Checks for Curvature (Four Dose Levels)

It is possible to use balanced designs incorporating more than three dose levels. Such designs are not commonly used as they have no advantage over simpler designs in routine work. They do, however, permit invalidity checks such as described in Section 2.8, over a wider range.

The calculation procedures for multidose level designs present some interesting features, which are exemplified by theoretical consideration of a four dose level assay.

It has been seen that in the three dose level assay, the response to the middose makes no contribution to measurement of the slope. It seems reasonable then that in the four dose level assay the second and third dose levels should make a smaller contribution than do the first and fourth. When the variance of response is independent of the value of the mean response to a particular dose, then a simple weighting procedure takes into account the varying precisions of the different contributions to the overall measure of slope.

The smaller contribution of the intermediate responses may be illustrated thus,

In Fig. 2.11 the horizontal distance d represents some measure of the dispersion of observations and is assumed to be the same at dose levels 1, 2, 3, and 4.

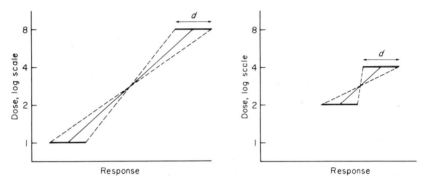

Fig. 2.11. Illustration of the varying reliability of values for slope when estimated over different ranges. The true slopes are indicated by the solid lines. The distance d represents a measure of dispersion of responses to a single dose (e.g., $\pm$ one standard deviation). If d is the same for all dose levels, then the probability of the estimated slopes lying between the extremes shown by the dashed lines is the same regardless of dose range. Clearly these extremes of estimated slope are closer to the true slope when the overall dose range is greater.

The dashed lines joining opposite ends of the horizontal lines illustrate the range of variation of the two estimates of slope for the same probability level. It is seen that the range of slope variation is three times as great over one dose interval than over three dose intervals. An estimate of slope over three dose intervals is therefore three times more reliable than an estimate over one dose interval. Thus in a four dose level assay the responses to extreme doses should be "weighted" by a factor of 3.

For example, d may be the range from -1 to $+1$ standard deviation. However, to demonstrate this point it is not necessary to define d or to state the probability of the extreme slopes shown by the dashed lines. The essential theoretical requirements are that d be the same for all levels (the condition of homoscedasticity) and therefore that the probability of the extreme slopes (shown by the dashed lines) be the same in both cases. In the agar diffusion assay, variance of response, and therefore standard deviation of response, is normally independent of dose level on the linear part of the log dose–response line. In fact, unless variances of responses to different dose levels differ greatly, then the weighting procedures to be described give a very good estimate of slope.

The purpose of Fig. 2.11 is to illustrate in a *qualitative* manner that a wider dose range leads to a more reliable estimate of slope and thus indicate the purpose of the weighting procedures to be described. It may also be inferred in a *qualitative* manner that in the case of a two dose level assay too, a wider dose ratio leads to a more reliable estimate of slope.

This must not be taken as an unqualified recommendation of wide overall dose ranges. It is shown in Table 8.3 that generally in agar diffusion assays the *precision* of the potency estimate is increased only slightly by overall dose ranges in excess of 2:1. Moreover, at dose ranges much in excess of 2:1, curvature of the dose–response line might lead to *bias* of the potency estimate.

The weighting procedure may now be explained by reference to Fig. 2.12, which represents a hypothetical log dose–response rectilinear relationship for doses of 1, 2, 4, and 8 units.

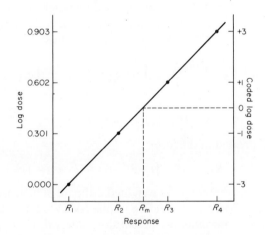

Fig. 2.12. An ideal log dose–response rectilinear relationship for doses of 1, 2, 4, and 8 units. Mean responses to the four dose levels are R_1, R_2, R_3, and R_4. The value R_m represents a hypothetical response to the geometric mean dose, which is at 0 on the coded log dose scale. This graph is explained in the text.

It is convenient to use "coded log doses" to avoid fractions and thus simplify the arithmetic. R_1, R_2, R_3, and R_4 represent the measured mean responses to the coded log doses -3, -1, $+1$, $+3$, and R_m represents the mean calculated response to the calculated mean coded log dose of 0.

The coding of the log doses is explained as follows:

(1) For convenience put the mean log dose $\bar{x} = 0$.

(2) Code the x's so that all x deviations can be expressed as integral distances in smallest units.

(3) The resulting code for four equidistant points is $-3, -1, +1, +3$, regardless of the original values and scale of the x's.

(4) In this case two coded units correspond to a log dose interval of 0.3010.

The weighting procedure is derived as follows. The four observed mean responses may be regarded as giving four separate estimates of slope. Taking the mean coded log dose 0 and its corresponding (hypothetical) response R_m as reference points:

(1) The four independent estimates of slope are

$$(R_m - R_1)/3, \quad (R_m - R_2)/1, \quad (R_3 - R_m)/1, \quad (R_4 - R_m)/3.$$

(2) It can be shown that each estimate should be weighted by the square of the coded x interval that it covers. The relative weights are then 9, 1, 1, 9. As the sum of these weights is 20, each individual weight must be divided by 20 to express the weighted mean in terms of a single coded unit.

(3) The resulting weighted estimate of slope is thus

$$\left(\frac{9}{20}\right)\left(\frac{R_m - R_1}{3}\right) + \left(\frac{1}{20}\right)\left(\frac{R_m - R_2}{1}\right) + \left(\frac{1}{20}\right)\left(\frac{R_3 - R_m}{1}\right) + \left(\frac{9}{20}\right)\left(\frac{R_4 - R_m}{3}\right)$$

which simplifies to

$$\tfrac{1}{20}[-3R_1 - R_2 + R_3 + 3R_4]$$

It can also be expressed in the form

$$(-3R_1 - R_2 + R_3 + 3R_4)/[(-3)^2 + (-1)^2 + (+1)^2 + (+3)^2]$$

These expressions represent the weighted estimate of increase in response over an increase in dose corresponding to one coded log dose unit. As there are two coded log dose units per actual dose interval, it follows that E is given by the expression

$$E = \tfrac{1}{10}[3R_4 + R_3 - R_2 - 3R_1]$$

However, if the response were measured as the sums of mean responses for sample and standard, thus doubling the number of individual estimates of slope, the fraction would revert to $\tfrac{1}{20}$ and the final expression would become

$$E = \tfrac{1}{20}[3(S_4 + T_4) + (S_3 + T_3) - (S_2 + T_2) - 3(S_1 + T_1)]$$

The use of these expressions for slope may be illustrated by application to the observations given in Example 1 for standard only:

$$E = \tfrac{1}{10}[3(19.40) + 17.25 - 15.28 - 3(13.33)]$$
$$= \tfrac{1}{10}[75.45 - 55.27] = 2.018$$

To express the slope in terms of a tenfold dose interval, divide by log dose interval, i.e., log 2,

$$b = E/I = 2.018/0.301 = 6.704$$

The calculation can be carried out without using coded log doses. Although the arithmetic is more tedious, it has the advantage of demonstrating the relationship between the factorial calculation just given and regression calculations, which are described later.

Table 2.5

Dose (IU/ml)	80	40	20	10
Log dose x	1.903	1.602	1.301	1.000
(Log dose) − (mean log dose)	+0.4515	+0.1515	−0.1505	−0.4515
Mean response (mm)	19.400	17.250	15.283	13.333
(Mean response) − (grand mean response)	3.085	0.935	−1.032	−2.982

$$\text{grand mean log dose } \bar{x} = 1.4515$$
$$\text{grand mean response } \bar{y} = 16.315 \text{ mm}$$

From the values given in Table 2.5, slope b is obtained without the use of coded log doses thus:

$$b = \frac{\begin{array}{c}(3.056 \times 0.4515) + (0.935 \times 0.1505) \\ + (-1.032 \times -0.1505) + (-2.982 \times -0.4515)\end{array}}{(0.4515)^2 + (0.1505)^2 + (-0.1505)^2 + (-0.4515)^2}$$

$$= 3.0353/0.4530 = 6.700$$

The similarity in principle to the previous calculation is made clear by expressing the former thus:

$$E = [3R_4 + R_3 - R_2 - 3R_1]/[3^2 + 1^2 + (-1)^2 + (-3)^2]$$

The calculation using log doses directly closely resembles the calculation of the "regression coefficient" but differs in that mean responses were used in place of individual responses.

The regression coefficient is defined by

$$b = S_{xy}/S_{xx} \tag{2.6}$$

where

$$S_{xx} = S(x - \bar{x})^2 \tag{2.7}$$

$$S_{xy} = S(x - \bar{x})(y - \bar{y}) \tag{2.8}$$

where S represents "the sum of," $\bar{x}$ (read as "x bar") is the mean of all x's, and $\bar{y}$ is the mean of all y's.

Rather than calculate means and differences from means, it is more convenient to use the expressions

$$S_{xx} = S(x - \bar{x})^2 = Sx^2 - (Sx)^2/n \tag{2.7}$$

$$S_{xy} = S(x - \bar{x})(y - \bar{y}) = Sxy - (Sx)(Sy)/n \tag{2.8}$$

where n is the total number of observations. The proof of this transformation is not given, but the reader may verify the identity by choosing a simple example and calculating both ways.

The regression coefficient for the doses and responses of Example 1 is calculated thus:

$$S_{xx} = 6[1.903^2 + 1.602^2 + 1.301^2 + 1.000^2]$$
$$- 6[1.903 + 1.602 + 1.301 + 1.000]^2/24$$
$$= 53.28248 - 50.56445 = 2.71803$$

$$S_{xy} = (1.903 \times 116.4) + (1.602 \times 103.5) + (1.301 \times 91.7) + (1.000 \times 80.0)$$
$$- [6(1.903 + 1.602 + 1.301 + 1.000)(116.4 + 103.5 + 91.7 + 80.0)]/24$$
$$= 586.618 - 568.407 = 18.211$$

$$b = 18.211/2.718 = 6.700$$

In contrast to the factorial method, the regression calculation is of relatively wide application. Apart from the straightforward case that has been demonstrated it may be used when (1) log dose intervals are unequal, and (2) the number of observations for each dose level is not constant. It is also applicable to direct dose–response relationships as described for slope ratio assays (Chapter 3).

The factorial method described is only applicable to parallel line assays when log dose ratios are evenly spaced, that is, when doses form a geometrical progression. The main virtue of the method is the ease of computation. The advantage of using a design to which the factorial method is applicable will now be clear to the reader. Factors for other designs of assay are given in Appendix 2.

2.10 Assays by Large Plates—General Principles

The variability of response to any one test solution both within a single plate and between plates was recognized in the early days of the diffusion assay. The causes of these variations and some remedies have been described in Section 2.2.

The use of large plates to minimize these variations was introduced by Brownlee *et al.* (1948) for streptomycin assays and (1949) for penicillin assays. These plates were constructed from plate glass and so in contrast to petri dishes were flat bottomed. This made it possible to obtain an agar medium layer of uniform thickness. Solutions were applied to the plate in accordance with a partially randomized 8 × 8 quasi-Latin square design. These designs were used for the streptomycin assays described in the first paper (three samples and one standard per plate) and also for the less precise penicillin assays, in which seven samples were assayed on each plate.

The addition of 64 test solutions one by one to a plate results in a considerable time difference between addition of the first and the last. This represents a difference in diffusion time and leads to bias in resultant zone sizes. Brownlee *et al.* describe a routine for application of test solutions designed to compensate for differences in diffusion time.

Lees and Tootill (1955a–c) describe developments in the use of large plates and give examples of a variety of designs for different purposes.

A 30 cm square plate will accommodate 64 zones in an 8 × 8 arrangement. This is a convenient number for comparing three samples and one standard, or seven samples and one standard at two dose levels or one sample and one standard at four dose levels. Alternatively, 36 zones in a 6 × 6 arrangement may be used. This is suitable for the comparison of two samples and one standard at two dose levels or one sample and one standard at three dose levels.

The designs may be randomized Latin squares or quasi-Latin squares, as shown in Figs. 2.13 and 2.14. Other such designs may be derived using the mathematical tables of Fisher and Yates (1963). In the case of a Latin square 8 × 8 assay for three samples and one standard at two dose levels, there are

5	6	8	7	3	4	1	2
3	1	7	6	2	8	5	4
6	8	3	5	4	1	2	7
8	3	2	4	1	7	6	5
1	4	5	2	7	6	3	8
7	5	6	3	8	2	4	1
4	2	1	8	5	3	7	6
2	7	4	1	6	5	8	3

Fig. 2.13. An example of a randomized Latin square design. Each of the numbers 1 to 8 occurs once and once only in each row and each column of the design.

A	I	7	10	3	12	14	16	5
B	15	9	8	13	6	4	2	II
A	5	3	14	7	16	10	12	I
B	4	6	II	2	9	15	13	8
A	10	16	I	12	3	5	7	14
A	14	12	5	16	7	I	3	10
B	8	2	15	6	13	II	9	4
B	II	13	4	9	2	8	6	15
	A	B	A	B	B	A	B	A

Fig. 2.14.   An example of a randomized quasi-Latin square design. In such a design, each of the 16 numbers occurs once and only once in four of the eight rows and four of the eight columns. There are two sets of rows and two sets of columns, which are distinguished here by the letters A and B. In A rows, the numbers are 1, 3, 5, 7, 10, 12, 14, and 16, in B rows, the numbers are 2, 4, 6, 8, 9, 11, 13, and 15, and in A columns, the numbers are 1, 4, 5, 8, 10, 11, 14, and 15, and in B columns, the numbers are 2, 3, 6, 7, 9, 12, 13, and 16. The consequence of such an arrangement is discussed in Section 6.10.

for each sample eight high dose and eight low dose zones. Each set of eight zones is distributed so that each zone appears once in each row and column. In the case of a quasi-Latin Square for seven samples and one standard, there are four high dose and four low dose zones for each sample. The design is such that either a high or a low dose zone for every sample occurs once in each row and once in each column.

When reading the zone sizes of a petri dish assay, the operator often knows to which solution the zone he is reading corresponds. In contrast to this, when reading a large plate the solutions have been applied in a randomized manner and he is not aware of which solution the zone corresponds to. Thus he is able to make a completely objective estimate of the zone size.

It is usual for one person to read the zones starting in the top left-hand corner, working along each row in turn. Results are called out to a second person who, using the Latin square design, writes the zone diameter directly in the appropriate column of the record sheet.

In accordance with the factors affecting zone size previously enumerated, large plates display some special features.

In applying the solutions to the plate, one correct practice is to start in the top left-hand corner, working along the row, then proceeding to the left-hand side of the second row, and continuing in this way finishing with the right-hand side of the bottom row. On average the solutions in the left-hand columns have a greater diffusion time than those in the right-hand column.

Similarly the solutions in the upper rows have a greater diffusion time than those in the lower rows. Thus there is a tendency for decreasing zone size from left to right and from top to bottom.

If application of test solutions proceeds at a constant rate using the routine described above, there is exact compensation for differences in diffusion time. This may be demonstrated by reference to the Latin square design of Fig. 2.13.

Allowing one arbitrary time unit for application of each test solution, the times of addition of the eight replicates of test solution 5 are

$$0, \quad 14, \quad 19, \quad 31, \quad 34, \quad 41, \quad 52, \quad 61, \qquad \text{total} = 252 \text{ units}$$

Considering now test solution 4, the times are

$$5, \quad 15, \quad 20, \quad 27, \quad 33, \quad 46, \quad 48, \quad 58, \qquad \text{total} = 252 \text{ units}$$

In the case of quasi-Latin squares, although compensation is not achieved for each test solution, it is achieved perfectly for each preparation, as shown by taking pairs of test solutions corresponding to a single preparation.

Similarly it can be shown that when using quasi-Latin squares, the average diffusion times for high and low dose test solutions are identical when the rate of application is uniform. Thus overall time effects do not lead to apparent deviations from parallelism.

According to Lees and Tootill (1955a) there is a further positional influence on zone size in large plates. This is a difference between inner zones and those in the rows and columns at the edge of the plate. Each inner zone is surrounded on all sides by reservoirs to which standard or test solutions have been applied. Water diffusing from these points keeps the moisture content in the center of the plate relatively higher than at the edge, where there is a tendency for the agar to dry out. Thus at the edges if the moisture content is lower, diffusion is faster and zones tend to be larger. These effects are to some extent counteracted by the use of a randomized design. However, loss of moisture suggests that lids are not well fitting, and therefore that practical steps may be taken to minimize this source of error. If plates are constructed with edges of aluminum, the increased rate of heat transfer at the edges may lead to more rapid commencement of growth, with a consequent edge effect in the opposite direction to that first described.

As an additional safeguard against bias, duplicate determinations employ different randomized Latin (or quasi-Latin) square designs. It is convenient in a busy laboratory to have available about 20 different designs of each type. These should each be in the form of a template that is placed beneath the large agar plate. The numbers are visible through the agar and correspond to the positions of the reservoirs, thus facilitating the process of applying test solutions in the correct positions.

In the case of assays using the 6 × 6 Latin square designs, potency calculations are identical with the corresponding assays using petri dishes (Examples 3 and 4). In the statistical evaluation described in Chapter 6 there is one additional feature characteristic of the design.

The calculation of potency for the 8 × 8 designs is the same in principle, but different factors are needed outside the brackets in the expressions for E and F according to the number of samples and number of dose levels. The calculations are shown in Examples 5 and 6.

Other Latin square designs that have been used in agar diffusion assays include: 4 × 4, for comparison of one sample with a standard at two dose levels; 9 × 9, for comparison of two samples with a standard at three dose levels. This design is described in the European Pharmacopeia, Volume II.

However, most routine work can be accommodated conveniently in the 6 × 6 and 8 × 8 designs, and a multiplicity of designs in any one laboratory has little to commend it.

For high-precision assays, 12 × 12 designs are described by Lees and Tootill (1955b). One of these includes two separate weighings and sets of dilutions to three dose levels each, for one sample and one standard; i.e., 12 treatments in all, which then occupy each row and each column of the design. The main application of these designs is in the establishment of reference standards.

2.11   Large Plate Assays Using Latin Square Designs

Example 5 refers to the assay of three samples of neomycin by an 8 × 8 randomized Latin square design. The actual design used was that shown in Fig. 2.13. Standard dose levels were 60 and 15 U/ml; samples were also diluted to nominal levels of 60 and 15 U/ml according to their expected potencies. Reference numbers were allocated to the eight test solutions and then these were applied to the agar plate in accordance with the Latin square design (odd numbers corresponding to high dose level).

Example 5:   Assay of Neomycin

Test organism: *Staphylococcus aureus* ATCC 6538-P
Dose ratio: 4:1
Standard: Neomycin sulfate, potency 646 U/mg,

$$187.0 \text{ mg} \rightarrow 200 \text{ ml} : 10 \text{ ml} \rightarrow 100 \text{ ml}$$

Samples: Experimental samples of neomycin sulfates:

(1)   217.8 mg → 200 ml : 10 ml → 100 ml

(2)   215.3 mg → 200 ml : 10 ml → 100 ml

(3)   98.8 mg → 250 ml : 25 ml → 100 ml

These dilutions of standard and sample gave nominal high level doses of 60 U/ml. They were further diluted to give low dose levels of nominal potency of 15 U/ml.

Test solutions were applied to wells in the agar medium using the same dropping pipette for all solutions and placing three drops in each well. Responses are shown in Table 2.6.

Table 2.6

Responses (zone diameter in mm) in the assay of neomycin (Example 5)[a]

	Sample 1		Sample 2		Sample 3		Standard	
	1	2	3	4	5	6	7	8
	23.3	21.1	22.7	19.6	23.3	21.0	23.1	20.5
	23.4	20.2	22.9	20.1	22.9	20.5	22.9	20.1
	22.9	20.1	22.6	19.5	22.9	20.9	23.0	20.3
	23.1	20.6	22.5	19.6	23.2	20.7	22.7	20.5
	23.6	20.3	22.8	20.3	23.6	20.9	23.0	20.4
	23.4	21.1	23.1	20.6	24.0	21.4	23.6	20.9
	23.9	21.3	23.8	20.5	24.0	20.4	23.7	20.9
	24.1	21.4	23.4	20.7	24.5	22.0	24.0	21.4
Treatment total:	187.7	166.1	183.8	160.9	188.4	167.8	186.0	165.0
Treatment mean:	23.463	20.763	22.975	20.113	23.550	20.975	23.250	20.625

[a] Test solutions are numbered 1 to 8; odd numbers are high doses, even numbers are low doses.

Difference due to dose:

$$E = \tfrac{1}{4}[(T_2 + T_2' + T_2'' + S_2) - (T_1 + T_1' + T_1'' + S_1)]$$
$$= \tfrac{1}{4}[(23.463 + 22.975 + 23.550 + 23.250) - (20.763 + 20.113 + 20.975 + 20.625)]$$
$$= \tfrac{1}{4}(93.238 - 82.476) = 2.6905$$

Slope:

$$b = E/I = 2.6905/0.6021 = 4.4685$$

Sample 1:

Difference due to sample:

$$F = \tfrac{1}{2}[(T_2 + T_1) - (S_2 + S_1)]$$
$$= \tfrac{1}{2}[(23.463 + 20.763) - (23.250 + 20.625)]$$
$$= \tfrac{1}{2}(44.226 - 43.857) = 0.1775$$

Log of potency ratio:

$$M = F/b = 0.1775/4.4685 = 0.03927$$

Potency ratio:

$$\text{antilog } 0.0393 = 1.095$$

Standard high dose test solution potency: 60.4 U/ml
Sample potency:

$$(60.4 \times 1.095 \times 100 \times 200)/(10 \times 217.8) = 607 \text{ U/mg}$$

Sample 2:

$$F' = -0.3936$$
$$M' = -0.08806 = \bar{1}.91194$$

Potency ratio:

$$\text{antilog } \bar{1}.9119 = 0.8164$$

Sample potency:

$$(60.4 \times 0.8164 \times 100 \times 200)/(10 \times 215.3) = 458 \text{ U/mg}$$

Sample 3:

$$F'' = 0.3250$$
$$M'' = 0.07273$$

Potency ratio:

$$\text{antilog } 0.0727 = 1.182$$

Sample potency:

$$(60.4 \times 1.182 \times 100 \times 250)/(25 \times 98.8) = 723 \text{ U/mg}$$

Sample potency ratios may alternatively be calculated simply by means of the tables (Appendix 3) using zone size totals (treatment totals) and the relationship

$$\frac{F}{E} = \frac{D/16}{B/32} = \frac{D}{B/2}$$

where D = (sum of individual sample responses) − (sum of standard responses)   and
B = (sum of all high dose responses) − (sum of all low dose responses).

2.12   Large Plate Assays Using Quasi-Latin Square Designs

For the simultaneous assay of larger numbers of samples, an 8×8 quasi-Latin square may be used. In this, seven samples are compared with one standard as shown in Example 6, the assay of vitamin B_{12}. The quasi-latin square design used in this example is shown in Fig. 2.14. Due to the reduced number of zones for each sample, the precision obtained by this design is less than that possible when using an 8×8 Latin square for three samples.

In this assay of vitamin B_{12}, the response is linear over a very wide range of log dose. The slope, however, is relatively small, about 4 mm for a tenfold increase in dose. These circumstances permit the use of a wide dose ratio. A ratio of $10:1$ is commonly used.

As already explained in Section 2.9, the use of a wider ratio gives a more reliable estimate of slope. However, it will be demonstrated in Chapter 8 that as in the agar diffusion method random error is generally very small compared with slope, measurement of slope has only little influence on precision of the assay. To summarize, the use of a wider dose ratio in the range of linear response results in a real but small improvement in precision.

Example 6:   Assay of Vitamin B_{12}

Test organism: *Escherichia coli* NCTC 8878
Design: 8×8 quasi-Latin square
Dose ratio: $10:1$
Standard: A solution of cyanocobalamin standardized spectrophotometrically so as to contain 10 $\mu g/ml$ of anhydrous cyanocobalamin was diluted to a high dose concentration of 0.2 $\mu g/ml$.
Samples: Liquid samples from various stages in the fermentation and extraction processes in the manufacture of vitamin B_{12}. Dilutions were all 1 in 50. Test solutions were added to wells cut in the inoculated medium at the rate of three drops per well. Observations are recorded in Table 2.7.

This example illustrates the calculation routines that are appropriate in a busy laboratory. Response totals are used instead of means and potency ratio is obtained very simply from F/E ratios.

Calculation procedure: From the tabulated zone diameters (Table 2.7) obtain:

(1)   The totals of the four responses to each treatment.
(2)   The totals of the eight responses for each of the seven samples $S(T^i)$ and the one standard $S(S)$.
(3)   The totals of all 32 responses to the high doses, $S(H)$.
(4)   The totals of all 32 responses to the low doses, $S(L)$.

Calculate

$$B = S(H) - S(L).$$

Table 2.7

The data and calculations of the assay of vitamin B_{12}, Example 6

	Sample/dilution to high dose level/test solution number														Standard (μg/ml)		
	A		B		C		D		E		F		G			0.2	0.02	
	1/50		1/50		1/50		1/50		1/50		1/50		1/50					
	1	2	3	4	5	6	7	8	9	10	11	12	13	14	15	16	
	24.7	20.0	24.3	20.1	24.2	19.5	24.4	19.5	25.3	20.3	25.3	20.8	25.3	21.0	24.9	20.8	
	24.6	20.1	24.1	19.9	24.2	19.3	23.9	19.8	24.6	20.6	25.1	20.4	25.1	21.1	24.3	20.5	
	24.4	19.7	24.2	19.8	23.9	19.2	24.4	19.9	25.4	20.6	24.7	20.6	25.2	20.8	24.6	20.4	
	24.4	19.8	24.1	19.6	23.7	19.3	24.8	19.2	24.9	20.1	24.6	20.7	24.9	20.6	24.7	19.9	
	98.1		96.7		96.0		97.5		100.2		99.7		100.5		98.5		787.2 = total of all high dose responses
		79.6		79.4		77.3		78.4		81.6		82.5		83.5		81.6	643.9 = total of all low dose responses
	177.7		176.1		173.3		175.9		181.8		182.2		184.0		180.1		Preparation total responses
	−2.4		−4.0		−6.8		−4.2		+1.7		+2.1		+3.9		—		D
	−0.0670 $\bar{1}.9330$		−0.1140 $\bar{1}.8860$		−0.1895 $\bar{1}.8105$		−0.1171 $\bar{1}.8829$		+0.0474		+0.0586		+0.7089				$M = D\frac{1}{4}B$
	0.857		0.769		0.647		0.764		1.115		1.145		1.285				Potency ratio
	8.6		7.7		6.5		7.6		11.2		11.5		12.9				Potency μg/ml

and then the seven values for the seven samples for

$$D = S(T^i) - S(S)$$

As B represents the sum of 32 differences and each value of D represents the sum of eight differences, it is seen that

$$B = 32E, \qquad D = 8F$$

It follows that

$$F/E = 4D/B$$

As the dose ratio is 10:1,

$$I = \log 10 = 1.0$$

so that

$$b = E/I = E$$
$$M = F/b = F/E = 4D/B$$

Thus for a 10:1 dose ratio assay, no special F/E table is required and potency ratios are obtained directly as antilog $4D/B$, or more conveniently as antilog $D/\frac{1}{4}(B)$.
   Substituting the observed values:

$$B = 787.2 - 643.9 = 143.3, \qquad \tfrac{1}{4}(B) = 35.825$$

Sample A:

$$D = 177.7 - 180.1 = -2.4$$
$$M = -2.4/35.825 = -0.0670 = \bar{1}.9330$$

Potency ratio of test solutions:

$$\text{antilog } \bar{1}.9330 = 0.857$$

Potency of high dose test solution:

$$0.857 \times 0.2 = 0.1714 \; \mu g/ml$$

Sample potency:

$$0.1714 \times 50 = 8.6 \; \mu g/ml$$

   Potencies of the remaining samples were obtained in the same way. They are shown in tabular form in Table 2.7.

2.13   Low-Precision Assays Using Large Plates

   A design suitable for the assay of even larger numbers of samples but giving results of quite low precision is illustrated by the large plate assay for bacitracin in Example 7. In this case the samples were submitted by a production plant as routine checks on various stages in the extraction and purification processes in bacitracin manufacture.

In such work a supposed potency is quoted in order to enable the analyst to make appropriate dilutions. The supposed potency may be a very rough guess and so it is desirable to have a wide range of standards with which to compare sample test solutions. A four dose level curve is employed covering an overall 8:1 range in steps of 2:1. Sample test solutions are prepared at a single level only, which is intended to lie within the range of the standard curve.

For rough work such as this, it suffices to draw a graph, usually the best straight line fitting all four points. However, so that the design may be evaluated and its limitations demonstrated, it is necessary to show a non-graphical computation.

To obtain E, the expression derived in Section 2.9 is used:

$$E = \tfrac{1}{10}[3S_4 + S_3 - S_2 - 3S_1]$$

The differences between samples and standard are given by

$$F = T_i - \tfrac{1}{4}[S_1 + S_2 + S_3 + S_4]$$

where T_i is the mean response for each sample in turn.

Assuming that samples are not diluted to the geometric mean of the standard doses (arithmetic mean on a log scale), potency estimate of sample test solution is calculated by equating the log of the ratio (estimated potency/nominal potency) to $M + \bar{x}_S - x_T$, where $\bar{x}_S$ is the mean of the logarithms of standard doses and x_T the logarithm of the sample nominal dose.

In this type of work the test solution potency may even be so different from supposed potency as to be outside the standard range and on a non-rectilinear part of the response line. It is for this reason, in contrast to previously described designs, that only a single dose level for each sample is used, as it is more satisfactory to base the estimation of slope on the standard response only. In this way, an occasional erratic sample whose response lies outside the range of linearity does not contribute to the determination of slope with consequent distortion of the response line.

There are no requirements regarding position of test solutions on the plate in this assay, and completely random distribution has been proposed. A random design, however, frequently results in the same solution appearing in adjacent positions, and so if zones are very large there might be overlapping of edges. It seems therefore that the use of a quasi-Latin square has some advantage.

If random distribution is used, the pattern should be based on a genuine random arrangement, as may be obtained by use of tables of random numbers such as those of Fisher and Yates (1963).

Example 7: Assay of Bacitracin

Test organism: *Micrococcus flavus* NCTC 7743

Design: 8 × 8 random distribution

Standard: Bacitracin working standard was diluted so as to give test solutions of potencies 0.5, 1.0, 2.0, and 4.0 IU/ml.

Samples: Liquid samples were diluted to a nominal value of 2.0 IU/ml. Test solutions were added to wells in the inoculated agar medium at the rate of 3 drops per well. Dilutions of samples and responses to all test solutions are shown in Table 2.8.

Calculation of potencies: The full calculation is given only for samples 9 and 7, whose responses are respectively close to and far from those expected from their supposed potencies. These data are processed further in Example 7e of Chapter 6 to contrast the influence of small and large values of M on the precision of the potency estimate.

From the standard responses:

$$E = \tfrac{1}{10}[3(23.38) + (21.10) - (19.28) - 3(17.23)] = 2.027$$
$$b = E/I = 2.027/0.3010 = 6.7343$$

For Sample 9:

$$F = 19.93 - \tfrac{1}{4}[17.23 + 19.28 + 21.10 + 23.38] = -0.32$$

As means of standard dose and nominal sample dose are unequal, calculate the ratio of estimated to nominal potency for the sample test solution via Eq. (2.5b), with

$$M = -0.32/6.7343 = -0.0476$$
$$\bar{x}_S = \tfrac{1}{4}(\log 4 + \log 2 + \log 1 + \log 0.5) = 0.1505$$
$$x_T = \log 2 = 0.3010$$

Thus,

$$M' = -0.0476 + 0.1505 - 0.3010 = -0.1980 = \bar{1}.8020$$

The ratio (R') of sample solution potency to its nominal potency is obtained as antilog $\bar{1}.8020$, that is, 0.6339. Sample potency is then calculated by

$$(\text{nominal potency}) \times R' \times (\text{dilution})$$

as

$$2.0 \times 0.6339 \times 10 = 12.7 \text{ IU/ml}$$

Similarly, for Sample 7,

$$F = 13.33 - 20.25 = -6.92$$
$$M = -6.92/6.7343 = -1.028$$
$$M' = -1.028 + 0.1505 - 0.3010 = -1.1785 = \bar{2}.8215$$
$$R' = \text{antilog } \bar{2}.8215 = 0.0663$$

Table 2.8

Data from the Assay of Bacitracin (Example 7)

	Sample												Standard			
Dilution:	1/20	1/20	1/20	1/20	1/1	1/1	1/1	1/1	1/10	1/10	1/10	1/10	Potency (IU/ml):	0.5	1.0	2.0	4.0
Test solution number:	1	2	3	4	5	6	7	8	9	10	11	12	Test solution number:	13	14	15	16
	22.1	21.0	21.3	20.1	14.3	—	13.3	—	20.1	19.1	19.2	18.2		17.0	19.1	21.2	23.9
	22.1	21.5	20.8	20.1	14.9	—	12.8	—	19.7	18.6	19.0	18.1		17.2	19.3	20.9	23.3
	22.7	20.8	20.9	20.9	13.9	—	12.9	—	19.8	18.8	18.8	18.3		16.8	19.5	21.2	23.2
	21.9	22.1	21.8	21.9	14.2	—	13.5	—	20.1	18.5	18.4	18.2		17.9	19.2	21.1	23.1
Total:	88.8	85.4	84.8	83.0	57.3	—	52.5	—	79.7	75.0	75.4	72.8		68.9	77.1	84.4	93.5
Mean:	22.20	21.35	21.20	20.75	14.33	—	13.33	—	19.93	18.15	18.76	18.20		17.23	19.28	21.10	23.38

which when multiplied by the nominal potency of the test solution (which was the undiluted sample) gives the estimated potency as

$$2.0 \times 0.0663 = 0.133 \text{ IU/ml}$$

This design of assay is only intended to give approximate results. In the case of figures that are effectively derived from an extrapolation of the standard range, the estimates obtained may be very rough indeed, since there is no guarantee that a linear extrapolation is valid. The extent of extrapolation in the case of Sample 7 may be seen on inspection of Fig. 2.15.

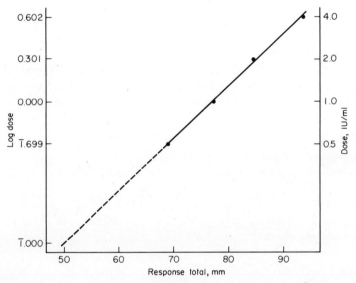

Fig. 2.15. A log dose–response curve for bacitracin using *Micrococcus flavus* as test organism. Mean responses to standard test solution from Example 7 are plotted versus log dose. The graph demonstrates the extent of extrapolation in the case of Sample 7 that gave the lowest response.

2.14 Small Plate Assays Using Interpolation from a Standard Curve

An unbalanced design incorporating a five-point standard curve is described in the US Code of Federal Regulations and has been in use for many years. Although widely used in some countries it is virtually unknown in Britain.

In this design each plate includes two treatments in triplicate. These appear in alternate positions on the plate, as shown in Appendix 1. In all plates one of the treatments is the reference dose, which is in fact the middose of the standard curve; the other treatment is either one of the standards of dose level number 1, 2, 4, or 5, or a sample of unknown potency at a dose level nominally the same as the standard reference point.

In contrast to the two and three dose level balanced assays previously described, each plate cannot be regarded as a self-contained assay and so allowance must be made for deviations in response attributable to the different plates. This is achieved by means of a "plate correction" based on the varying mean responses to the constant reference dose. A better term would be "plate group correction" since the calculated reference point is the mean of nine responses to reference dose in a group of three replicate plates. The simplifying assumption is made that if the three replicate plates are handled together at the same time, intragroup differences will be small and a common correction will suffice.

The method is illustrated by the assay of kanamycin in Example 8, which shows only one unknown sample. In routine practice, of course, many samples can be assayed at the same time with one standard curve.

Example 8:   The Assay of Kanamycin by an FDA Method

Test organism: *Bacillus pumilis* NCTC 8241
Standard: Kanamycin of potency 812 IU/mg, diluted as follows:

$$31.5 \text{ mg} \rightarrow 25.6 \text{ ml} : 2 \text{ ml} \rightarrow 200 \text{ ml}$$

giving a solution of potency 10 IU/ml, which was further diluted to give test solutions for the standard curve thus:

$$3.2 \text{ ml} \rightarrow 10 \text{ ml}$$
$$4.0 \text{ ml} \rightarrow 10 \text{ ml}$$
$$5.0 \text{ ml} \rightarrow 10 \text{ ml}$$
$$6.25 \text{ ml} \rightarrow 10 \text{ ml}$$
$$7.81 \text{ ml} \rightarrow 10 \text{ ml}$$

Sample: Kanamycin injection of labeled potency 1 g kanamycin base per vial. The average weight of 10 vials was found to be 1.284 g.

A portion was weighed and diluted thus:

$$246.1 \text{ mg} \rightarrow 200 \text{ ml} : 5 \text{ ml} \rightarrow 100 \text{ ml} : 10 \text{ ml} \rightarrow 100 \text{ ml}$$

giving a test solution of nominal potency 5.0 IU/ml. Test solutions were added by dropping pipette to stainless steel cylinders placed on the surface of the agar medium filling them almost to the brim. Responses are shown in Table 2.9.

The computation of sample potency is shown first by the graphical method of the FDA. In the interests of uniformity the symbols used here are not those of the FDA

Table 2.9

Observations—Zone Diameter (mm)

	Standards								Sample	
	3.2	R	4.0	R	6.25	R	7.81	R	U	R
	14.6	16.1	14.7	15.8	16.6	15.6	17.3	15.6	15.3	15.7
	14.1	15.6	15.1	15.6	16.8	15.8	17.0	15.6	15.8	15.8
	13.8	15.8	14.8	15.5	16.3	16.0	17.0	15.5	15.7	15.7
	14.5	16.0	14.7	15.7	16.6	15.8	17.3	15.6	15.8	15.9
	14.1	15.9	14.9	15.5	16.5	15.6	17.4	15.7	15.8	15.7
	14.4	16.2	15.2	15.6	16.2	15.7	17.2	15.5	15.5	15.7
	14.0	15.7	14.8	15.7	16.9	16.1	17.3	15.9	15.2	15.5
	14.2	15.7	15.0	15.4	16.5	15.7	17.3	15.8	15.1	15.8
	14.1	15.8	14.3	15.3	16.8	15.8	16.7	15.8	15.1	15.3
Total	127.8	142.8	133.5	140.1	149.2	142.1	154.5	141.0	139.3	141.1
Mean	14.20	15.87	14.83	15.57	16.58	15.79	17.17	15.67	15.48	15.68
Correction	−0.15		+0.15		−0.07		+0.05			
Corrected mean	14.05		14.98		16.51		17.22			

but an extension of the system used in the International and European Pharmacopeias. The subscript c is added to denote *corrected* mean responses.

The Code of Federal Regulations directs that if the five corrected mean responses plotted on an arithmetic scale against doses on a logarithmic scale appear to describe a straight line, then the two following values be calculated. These values may be described as *corrected ideal* responses to low and high doses. They are denoted by S_{1m} and S_{5m}, respectively, and take into consideration first the plate corrections and second the contributions of the responses at dose levels 1, 2, 4, and 5 to the slope.

The two values are calculated as

$$S_{1m} = \tfrac{1}{5}[3S_{1c} + 2S_{2c} + S_{3c} - S_{5c}]$$
$$S_{5m} = \tfrac{1}{5}[3S_{5c} + 2S_{4c} + S_{3c} - S_{1c}]$$

Substituting the actual values,

$$S_{1c} = 14.05, \qquad S_{2c} = 14.98$$
$$S_{3c} \text{ (the reference point)} = 15.72$$
$$S_{4c} = 16.51, \qquad S_{5c} = 17.22$$
$$S_{1m} = \tfrac{1}{5}[(3 \times 14.05) + (2 \times 14.98) + (15.72) - (17.22)] = 14.12$$
$$S_{5m} = \tfrac{1}{5}[(3 \times 17.22) + (2 \times 16.51) + (15.72) - (14.05)] = 17.27$$

These two values for S_{1m} and S_{5m} are plotted on semilog graph paper and joined to give the best straight line that can be drawn from the experimental data. This is illustrated in Fig. 2.16.

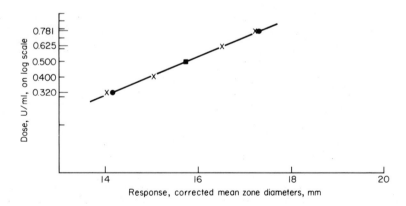

Fig. 2.16. The assay of kanamycin by 5 + 1 design using *Bacillus pumilis* as test organism (Example 8). This graph was drawn on two-cycle general purpose semilog/arithmetic graph paper such as commonly used in this work. Note the narrow range of the log scale into which all observations are compressed. The resultant difficulty of reading the graph reduces the capability of the assay method. ×, mean responses to standard test solutions after correcting for plate differences; ■, reference point; ●, calculated ideal responses to high and low dose.

It is seen that on the set of three plates for the sample, the mean of the nine sample responses (15.48 mm) is 0.20 mm less than the mean of the nine responses to the reference dose on the same set. The sample test solution potency is then obtained by reading from the graph the dose that corresponds to a response 0.20 mm to the left of the intersection of the dose–response line with the horizontal line at dose level 5.0 IU/ml. It is found to be 4.72 IU/ml.

Taking into consideration weighings and dilutions, the kanamycin potency is calculated as

$$\frac{4.72 \times 100 \times 100 \times 200}{10 \times 5 \times 246.1} = 767 \text{ IU/mg}$$

It follows that a vial of average content of kanamycin sulfate (1.284 g) will contain

$$767 \times 1284 = 986,000 \text{ IU of activity}$$

The following alternative method of calculation avoids the tedium of reading from a graph and the resulting inaccuracy:

$$E = \tfrac{1}{10}[2S_{5c} + S_{4c} - S_{2c} - 2S_{1c}]$$
$$= \tfrac{1}{10}[2(17.22) + (16.51) - (14.98) - 2(14.05)] = 0.787$$
$$F = T^i - S_3{}^i = 15.48 - 15.68 = -0.20$$
$$b = E/I = 0.787/0.0969 = 8.123$$
$$M = F/b = -0.20/8.123 = -0.0246 = \bar{1}.9754$$

Potency ratio:

$$\text{antilog } \bar{1}.9754 = 0.9450$$

Taking into consideration the weighing and dilutions as before, sample potency is calculated as 768 IU/mg.

The relationship between the expressions for E on the one hand and S_{1m} and S_{5m} on the other is readily demonstrated.

The best estimate of difference in response over four dose intervals is $(S_{5m} - S_{1m})$. This is equal to

$$\tfrac{1}{5}[3S_{5c} + 2S_{4c} + S_{3c} - S_{1c}] - \tfrac{1}{5}[3S_{1c} + 2S_{2c} + S_{3c} - S_{5c}]$$

which simplifies to

$$\tfrac{1}{5}[4S_{5c} + 2S_{4c} - 2S_{2c} - 4S_{1c}]$$

It follows that the best estimate of difference in response over one dose interval is one quarter of this value, i.e.,

$$\tfrac{1}{10}[2S_{5c} + S_{4c} - S_{2c} - 2S_{1c}]$$

which is the expression used to calculate E for this assay design. Derivation of the expression for E is analogous to that shown for a four dose level assay in Section 2.9.

Certain features of this design are open to criticism.

(1)   The five dose levels are presumably intended to determine whether the response line of each individual assay is straight or curved. However, in routine assays as carried out in most laboratories it would not usually be possible to distinguish between curvature and random error. In practice it seems that results are computed invariably on the assumption of a straight line relationship, and so the five points are an unnecessary complication. Moreover, the necessity of the five doses forming a geometrical progression (for ease of computation) leads to awkward dilutions, which preclude the use of standard volumetric bulb pipettes. Although not an essential feature of the method, the measurement of volumes as small as 0.64 ml is a common malpractice. The facts of curvature should be known in advance in any routine method. There is no need to rediscover them in each individual test! The significance of curvature is discussed in Section 8.4.

(2)   The use of only a single dose level for the unknown sample precludes any check on the basic assumption of parallel response lines for standard and unknown. In the routine assay of a product of known composition it is not essential to establish the fact of parallelism on every occasion. In contrast, in those cases when full details as to the nature, formulation, and history of the sample are not known with certainty, then the absence of this simple and economical check is a disadvantage.

2.15   Missing Values

It sometimes happens that an observed response is apparently spurious. Possibly the zone is misshapen or perhaps it is so discrepant from other responses supposedly of the same treatment that quite clearly the wrong test solution has been applied. If and only if there are good grounds for believing that the observed response is not an extreme case of random variation but is spurious, then it should be replaced by a value calculated from the true observations. The United States Pharmacopeia describes tests for identification of outlying or aberrant observations. However, these are statistical tests based on probability and as such are not infallible. In a *micro*biological assay where cost of testing is small as compared with many *macro*biological assays and replication is often relatively high, it is perhaps safer to reject all observations in the block (e.g., plate) containing an outlier.

It might appear in the assay of streptomycin (Example 3) that the response (15.7 mm) to test solution 2 is an outlier. In fact the total of all responses on this plate is lower than all others and is 3.0 mm less than the mean of the other five plates. The experimental design automatically corrects for this discrepancy.

On those occasions when it seems right to replace the spurious observation, then a figure calculated by means appropriate to the assay design should be substituted.

If the assay is by a completely random design such as the assay of bacitracin (Example 7), then the missing value is replaced simply by the mean of the other responses to the same treatment. If, however, the assay is by a randomized block design in which each block may be regarded as a complete assay in itself, then a better estimate of what the missing value should have been is obtained by taking into consideration the differences between the blocks. Examples 1–4 are all randomized block designs in which each plate is a block.

In Latin square designs the process is taken a stage further by making allowance not only for rows (which we may say arbitrarily correspond to blocks), but also for columns.

The European and United States Pharmacopeias give formulas for calculating these replacement values. If two values are missing the correct procedure is to guess the value of one of these and then make a first estimate of the still missing value by use of the appropriate formula. Then using this estimate, calculate a value for the second missing observation to replace the guessed value. Using this estimate for the second, make a new estimate for the first missing observation. Continue this process until successive recalculations produce negligible differences in estimates of the replacement values.

Similar iterative calculations may be used if more than two responses are missing. Naturally, missing responses lead to less reliable results. Such estimations of replacement values do not add any information but merely restore the balance of the design, thus making possible calculation of potency estimate and confidence limits (see Chapter 6) by the normal procedures. In the calculation of confidence limits each replacement value results in the loss of one degree of freedom in the "error squares" and of course in the total degrees of freedom.

The calculation of replacement values is illustrated by two examples.

(1)   *A randomized block design.*   Suppose that in the assay of streptomycin (Example 3), the response to test solution 6 in the first plate (actually 15.3 mm) is missing:

B' = sum of all (5) responses in the block (plate) from which a value is missing = 91.8

T' = sum of all (5) responses to the treatment from which a value is missing = 80.5

G' = sum of all (35) responses recorded = 642.6

Calculate the replacement value:

$$y' = \frac{nB' + kT' - G'}{(n - 1)(k - 1)} \tag{2.9}$$

where k is the number of treatments (6) and n the number of replicates (6). Thus in this case

$$y' = \frac{(6 \times 91.8) + (6 \times 80.5) - (642.6)}{(6 - 1)(6 - 1)} = 15.6 \text{ mm}$$

(2) *A Latin square design.* Suppose that in the assay of neomycin (Example 5), the response to test solution 6 in the seventh row and eighth column of the design (actually 20.4 mm) is missing. Calculate values analogous to those in the previous illustration, where this time B' corresponds to a row (7) and a new value C' corresponds to a column (8).

$$B' = 158.1 \qquad C' = 154.6, \qquad T' = 147.4, \qquad G' = 1404.9$$

Calculate the replacement value:

$$y' = \frac{k(B' + C' + T') - 2G'}{(k - 1)(k - 2)} \tag{2.10}$$

In this design, $n = k = 8$, so that

$$y' = \frac{8(158.1 + 154.6 + 147.4) - (2 \times 1404.9)}{7 \times 6} = 20.7 \text{ mm}$$

These formulas are in fact those described by Emmens (1948) but with changed nomenclature.

References

Brownlee, K. A., Delves, C. S., Dorman, M., Green, C. A., Grenfell, E., Johnson, J. D. A., and Smith, N. (1948). *J. Gen. Microbiol.* **2**, 40.
Brownlee, K. A., Loraine, P. K., and Stephens, J. (1949). *J. Gen. Microbiol.* **3**, 347.
Bryant, M. C. (1968). "Antibiotics and Their Laboratory Control." Butterworths, London.
Cooper, K. E. (1963 and 1972). *In* "Analytical Microbiology" (F. W. Kavanagh, ed.), Vols. I and II. Academic Press, New York.
Cooper, K. E., and Gillespie, W. A. (1952). *J. Gen. Microbiol.* **7**, 1.
Cooper, K. E., and Linton, A. H. (1952). *J. Gen. Microbiol.* **7**, 8.
Cooper, K. E., and Woodman, D. (1946). *J. Pathol. Bacteriol.* **58**, 75.
Emmens, C. W. (1948). "Principles of Biological Assay." Chapman and Hall, London.
Fisher, R. A., and Yates, F. (1963). "Statistical Tables for use in Biological, Agricultural and Medical Research," 6th ed. Longman, London.
Humphrey, J. H., and Lightbown, J. W. (1952). *J. Gen. Microbiol.* **7**, 129.
Kavanagh, F. W. (1972a). Private communication.

Kavanagh, F. W. (1972b). "Analytical Microbiology," Vol. II. Academic Press, New York and London.

Kavanagh, F. W. (1974). *J. Pharm. Sci.* **63**, 1459.

Knudsen, L. F., and Randall, W. A. (1945). *J. Bact.* **50**, 187.

Lees, K. A., and Tootill, J. P. R. (1955a). *Analyst* **80**, 95.

Lees, K. A., and Tootill, J. P. R. (1955b). *Analyst* **80**, 112.

Lees, K. A., and Tootill, J. P. R. (1955c). *Analyst* **80**, 531.

Mitchison, D. A., and Spicer, C. C. (1949). *J. Gen. Microbiol.* **3**, 184.

CHAPTER 3

TUBE ASSAYS FOR
GROWTH-PROMOTING SUBSTANCES

3.1 General Principles

The practical details of a typical tube assay for growth-promoting substances have been outlined in Chapter 1.

The basis of estimation of potency is the relationship between applied dose of growth substance and the resulting growth of the test organism in an incomplete medium, i.e., a medium containing all except one of the ingredients essential for growth. The missing ingredient is that growth promoting substance which is to be assayed and which is supplied to individual tubes of medium as graded doses of standard or of sample.

The growth of the test organism results in the initially clear test preparation becoming turbid. This growth may be measured photometrically either in a simple nephelometer, which measures light scattered by the suspended cells, or by an absorptiometer, which measures the reduced transmittance as compared with a blank test preparation not inoculated with test organism.

In the case of acid-producing organisms, the liberated acid may be the response used. The acid can be titrated with 0.1 N sodium hydroxide solution and the titer for each tube recorded in milliliters, or alternatively the resulting change in pH may be measured.

It is important to consider all possible influences on the growth of the test organism. Appropriate steps may then be taken to minimize unwanted effects and also to provide conditions in which the only effective restriction on growth is the graded dose of growth-promoting substance that we wish to assay.

These influences may be summarized as:

(1) *Dose of the missing growth substance* (test substance), which is applied in the forms of standard (known) and sample (unknown) test solutions.

(2) *Period of incubation.* For the ideal assay the period of incubation would be long enough to permit complete utilization of the added g.p.s., thus giving a response directly proportional to dose. Periods as long as three days are used in assays where the response measured is dependent on acid production. Assays using an optical response commonly have an incubation period of only 18 hours. If the incubation period does not permit complete utilization of the added g.p.s., then responses to higher doses are propor-

tionately smaller, thus leading to curvature of the response line. Minimum incubation time should be ascertained for each kind of assay and each batch of prepared medium.

(3) *Inoculum—quantity and stage of growth.* In an assay with a long period of incubation these are not very critical factors.

(4) *Temperature of incubation.* Small differences up or down from the optimum have a very significant effect on growth rate. If the period of incubation were long enough to permit complete utilization of the added g.p.s. even at the lower growth rate, then small differences in temperature would be unimportant. At shorter incubation periods the growth rate influences the finally measured response and so temperature should be very carefully controlled. Ideal conditions of incubation are provided by uniform tubes immersed in a well-stirred water bath. This is discussed in greater detail in Section 4.1.

(5) *Nature of the synthetic medium.* An adequate excess of all growth substances other than the one to be measured is essential, for otherwise additional limiting factors are introduced; pH must be conducive to growth. It should be noted that manufacturers of synthetic media recommend mild heat treatment of the reconstituted media (e.g., 115°C for 10 min or 100°C for 20 min) to kill off extraneous microorganisms before inoculation with the test organism. Normal sterilization procedures are liable to reduce the capability of supporting growth.

(6) *Nature of the test organism.* So far as possible a test organism should be chosen that is selective in its growth requirements and will not respond to extraneous substances that may be present in the sample. When naturally occurring substances include several related components, their differing relative activities to different test organisms creates many problems. The complexity of these problems may be exemplified by vitamin B_{12} (Bessell and Shaw, 1960) and by folic acid (Bird and McGlohon, 1972).

(7) *Inhibiting substances.* These might possibly be present as adsorbed traces of antibiotics, detergents, etc., on glassware, or as ingredients of the sample. The importance of a rigorous routine for cleaning glassware used in g.p.s. assays cannot be overemphasized. Suitable procedures are described by Cooperman (1972), Kavanagh (1963), and Skeggs (1963).

Of all these factors, the first is the essential basis of the assay. Conditions should be so regulated as to ensure a good response to the first and also to minimize unwanted variation. Ideally the extent of growth should be a direct function of dose only and should not be limited by the other influences described.

The practical steps needed to attain, or at least approach the required response and to eliminate unwanted variation are very adequately discussed by Kavanagh (1963).

3.2   Measurement of Response

Response in tube assays (whether of g.p.s. or antibiotics) is most commonly measured in terms of optical properties of the resulting cell suspension.

The concentration of the cell suspension may be estimated in two ways:

(1)   by measurement of the intensity of the selected light band scattered by the suspended cells (nephelometry),

(2)   by measurement of the proportion of the selected light band that is transmitted through the suspension to a detector in the light path (absorptiometry).

Since the process is essentially one of scattering rather than absorbance the former principle seems the more appropriate. As will be shown later, it has some distinct practical advantages. Rather surprisingly, absorptiometric measurements appear to be widely used. It is interesting to speculate as to the reasons for the popularity of absorptiometric measurements. Possible reasons are:

(1)   The availability of absorptiometers that have already been purchased with colorimetric methods in mind.

(2)   Due to the empirical manner in which turbidimetric assay responses are so widely interpreted, consideration of the nature of the detection system has seemed irrelevant to most analysts.

Absorptiometric measurements of cell suspensions and instrumental factors are reviewed by Kavanagh (1963, 1972), who describes modifications to certain instruments so as to give a measured response (absorbance) more closely proportional to the true response, which is taken to be cell concentration. This assumption, however, is a deliberate simplification, as cell size may vary according to the age of the culture and the presence or absence of antibiotics, thus affecting optical properties.

Calibration curves may be prepared relating measured absorbance of a particular species of organism in suspension to cell concentration. This procedure is illustrated in Fig. 3.1 for two instruments. It might appear from inspection of these graphs that little error would arise if absorbance values up to about 0.2 were taken to be equivalent to cell concentration. However, it is shown in Section 8.8 that in slope ratio assays (see Section 3.4) even slight curvature might introduce substantial bias.

Kavanagh (1963) stresses the problems of flow birefringence in absorptiometric measurements of suspensions of rod-shaped organisms. Turbulence after pouring the suspension into the cuvette causes fluctuations in measurements until the organism has returned to its completely random orientation state. This may take up to 20 sec.

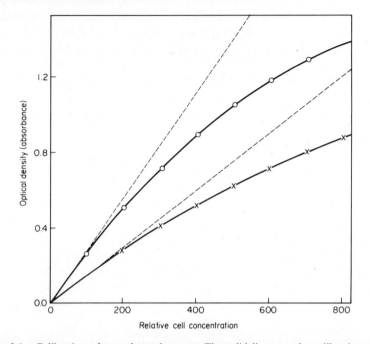

Fig. 3.1. Calibration of two absorptiometers. The solid lines are the calibration curves relating optical density and relative cell concentration. The dashed lines serve to show the extent of deviation from Beer's law. For routine transformation of absorbance to relative cell concentration it is more convenient to express the relationship in tabular form. ○, the spectronic 20 spectrophotometer; ×, the Coleman junior spectrophotometer.

The problem is overcome in the automated Autoturb® system, which employs a flow cell (A. H. Thomas and Co No. 9120–NO5) for optical measurements. The orientation of rod-shaped organisms under standard conditions of flow is sufficiently uniform to eliminate fluctuations in transmittance measurements.

In manual methods, measurement of scattered light may be made by means of a simple but convenient instrument, the EEL nephelometer. Using this instrument, incubation can be carried out in standard optically uniform tubes, which are later placed directly in the cuvette holder for measurement of scattered light.

It has been shown that for cell suspensions of *Staphylococcus aureus* of concentrations such as encountered in the normal working range for turbidimetric assays, the EEL nephelometer gives a measured response (arbitrary units) that is directly proportional to cell concentration.

The instrument responds to light scattered through a wide range of forward angles, and so flow birefringence would not be expected to present any problems. The author has not observed any of the difficulties ascribed to flow birefringence in the measurement of optical properties of rod-shaped organisms, which have been mentioned in the case of absorptiometric measurements.

In the case of photometric assays, variations in observed results also arise due to the following sources of error in optical measurements:

(1) Nonuniformity of tubes (if the measurements are made in the original tubes and not by transferring to a standard cell).

(2) Inadequate dispersion of the suspension.

(3) Variations in time between shaking to disperse the organism and measuring its optical properties; too short a time may cause interference from minute suspended air bubbles; too long a time may result in settling of the organism. Yeast cells settle rapidly.

The convenience of pH as a measure of response has already been mentioned. This has been used by Kavanagh (1973) for many years. He describes its application to response measurement in the 3 day assays for folic acid and vitamin B_{12} and defines the practical requirements for accurate assays as:

(1) pH meter must have an expanded scale so as to make precise measurements possible.

(2) Solution temperatures must not vary by more than $\pm 0.1°C$ during the period of pH measurement.

(3) The pH meter must be stable to at least ± 0.01 pH unit.

(4) Absolute value of pH is unimportant.

(5) Glass electrodes respond to changes in pH in milliseconds after being exposed to the solutions. Certain reference electrodes, especially those with fiber junctions, respond in seconds, thus causing a slow drift in measurements. These drifts are too large for accurate work. Reference electrodes with glass sleeve junctions respond rapidly. Combination electrodes may or may not be suitable; each should be tested for speed of response and stability.

3.3 Nonideal Responses

Ideally the response in g.p.s. assays is directly proportional to dose. In practice this ideal is often not achieved and so it is instructive to consider some actual responses.

Taking the case of assays in which the response used is an optical measurement of the resulting turbidity, some practically determined curves are shown in Fig. 3.2. In all these four assays by Önal (1971), response was measured using an EEL nephelometer and so instrument reading can be taken to be approximately proportional to cell concentration. It is seen that in the assay of methionine response is directly proportional to dose at lower

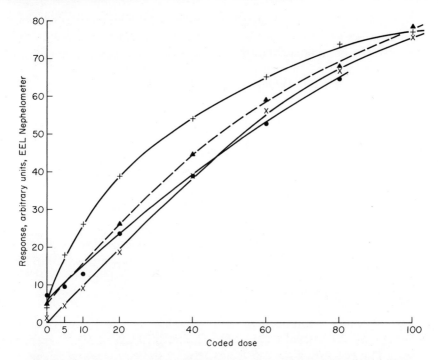

Fig. 3.2. Curvilinear responses in turbidimetric assays of growth-promoting substances. Turbidity was measured by EEL nephelometer and is recorded in arbitrary units. The response line for methionine is straight at lower doses. ×, methionine; +, thiamine; ●, pantothenic acid; ▲, nicotinic acid.

levels, whereas response lines for pantothenic acid, nicotinic acid, and thiamine show varying degrees of curvature.

To avoid the necessity to interpolate from a curve in such nonideal cases, Tsuji *et al.* (1967) derived a curve-straightening procedure. Tsuji measured bacterial growth by transmittance at 650 nm using a Beckman Model C colorimeter. He found that observations could be represented by the expression

$$T(x) = Ae^{-Bx} + C \qquad (3.1)$$

where x is the dose in concentration units, $T(x)$ the percentage transmittance at dose x, and A, B, C parameters to be estimated from the data.

For zero dose, substitution of $x = 0$ in Eq. (3.1) gives

$$T(0) = A + C \qquad (3.2)$$

Considering now a hypothetical infinite dose, substitution of $x = \infty$ gives

$$T(\infty) = C$$

This hypothetical infinite dose corresponds to the practical conditions when dose is so high that any further increase is without influence on the response. The limiting factor is then the ability of the medium to support growth in the presence of an excess of the substance to be determined.

It follows from Eq. (3.1) that

$$T(x) - C = Ae^{-Bx} \quad \text{and} \quad \ln[T(x) - C] = \ln A - Bx$$

and substituting $T(\infty) = C$,

$$\ln[T(x) - T(\infty)] = \ln A - Bx \tag{3.3}$$

Thus if the data can be represented by Eq. (3.1), a plot of $\ln[T(x) - T(\infty)]$ versus x results in a straight line with slope $-B$ and intercept $\ln A$ at zero dose.

To obtain this straight line relationship it is first necessary to evaluate $T(\infty)$. This might be done by including a tube with a very large excess of the g.p.s. to be determined, so that it is no longer a factor limiting growth. However, Tsuji (1973) estimates $T(\infty)$ from responses to three equally spaced doses as follows:

If the responses to three equally spaced doses x_1, x_2, and x_3 are expressed as $\ln[T(x_1) - T(\infty)]$, $\ln[T(x_2) - T(\infty)]$, and $\ln[T(x_3) - T(\infty)]$, respectively, a plot of these values versus dose gives a straight line (Fig. 3.3), from

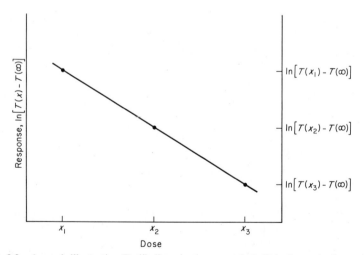

Fig. 3.3. A graph illustrating Tsuji's linearization procedure. It is shown in the main text that if Eq. (3.1) is a true representation of dose–response lines in turbidimetric assays of growth-promoting substances, then a plot of response in the form $\ln[T(x) - T(\infty)]$ versus dose would be a straight line. The graph is used in the derivation of Eq. (3.4) for the expression for $T(\infty)$.

which it is seen that

$$\ln[T(x_3) - T(\infty)] - \ln[T(x_2) - T(\infty)] = \ln[T(x_2) - T(\infty)] \\ - \ln[T(x_1) - T(\infty)]$$

By taking antilogs of both sides,

$$[T(x_3) - T(\infty)]/[T(x_2) - T(\infty)] = [T(x_2) - T(\infty)]/[T(x_1) - T(\infty)]$$

Rearranging this expression, $T(\infty)$ is evaluated as

$$T(\infty) = \frac{[T(x_2)]^2 - [T(x_1)][T(x_3)]}{2[T(x_2)] - [T(x_1) + T(x_3)]} \tag{3.4}$$

The effectiveness of this curve-straightening procedure is illustrated by the data of Tsuji for assays of vitamin B_{12}, calcium pantothenate, and pyridoxine. In Fig. 3.4, observations as originally recorded in percent transmittance are plotted versus coded log doses (coded so that all three assays

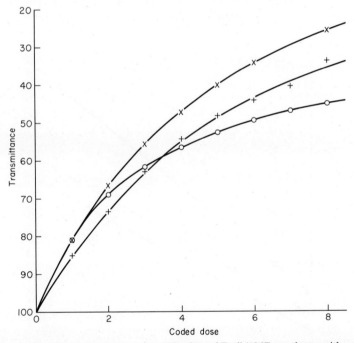

Fig. 3.4. Vitamin assay response curves. The data of Tsuji (1967) are shown with responses in their original form as percentage of transmittance. Doses are coded for convenience of presentation. Response curves are $\times$, vitamin B_{12} assayed using *Lactobacillus leichmannii*; $+$, calcium pantothenate assayed using *Saccharomyces carlsbergensis*; $\bigcirc$, pyridoxine assayed using *Saccharomyces carlsbergensis*.

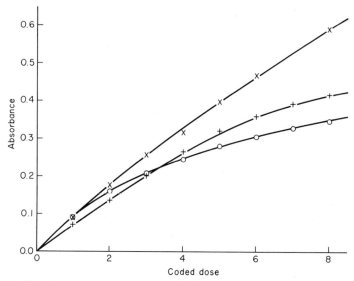

Fig. 3.5. Vitamin assay response curves. The data of Tsuji (1967) as used in Fig. 3.4 are shown with responses transformed to absorbance. Doses are coded for convenience of presentation. Response curves are ×, vitamin B_{12}; +, calcium pantothenate; ○, pyridoxine.

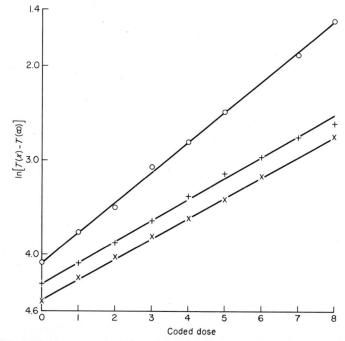

Fig. 3.6. Vitamin assay linearized response curves. These graphs represent the same data of Tsuji (1967) as used in Figs. 3.4 and 3.5. Responses have been expressed in the form $[T(x) - T(\infty)]$ and plotted as their natural logarithms. Linearized response lines are ×, vitamin B_{12} $(T(\infty) = 10)$; +, calcium pantothenate $(T(\infty) = 25)$; ○, pyridoxine $(T(\infty) = 40.5)$. The calculation of values of $[T(x) - T(\infty)]$ is illustrated for pyridoxine in Example 9.

may be shown on the same scale). In Fig. 3.5, the same data are converted to absorbance, which may be considered to approximate to the true response.

Values of $T(\infty)$ for the three assays were calculated using Eq. (3.4) and found to be: for vitamin B_{12}, 10; for calcium pantothenate, 25; and for pyridoxine, 40.5.

Using these values, responses were expressed in the form $[T(x) - T(\infty)]$ and plotted on a log scale in Fig. 3.6, in which the effectiveness of the linearization procedure is clearly demonstrated.

The method of calculation is demonstrated using Tsuji's data for pyridoxine in Example 9.

Example 9: Linearization of a Standard Curve for Pyridoxine

Test organism: *Saccharomyces carlsbergensis*
Response: Transmittance at 650 nm; Beckman model C colorimeter
Observations: See Table 3.1.

Table 3.1

Dose (x) ng/ml	% Transmittance
0.00	100.0
0.25	81.5
0.50	69.0
0.75	62.0
1.00	57.0
12.5	52.5
1.50	49.5
1.75	47.0
2.00	45.0

Substituting in Eq. (3.4) the values for responses to doses $x_1 = 0.00$, $x_2 = 1.00$, and $x_3 = 2.00$,

$$T(\infty) = \frac{(57.0)^2 - (100.0)(45.0)}{2(57.0) - (100.0 + 45.0)} = \frac{3249 - 4500}{114 - 145} = 40.5$$

Putting $T(\infty) = 40.5$, $\ln[T(x) - T(\infty)]$ is calculated for each dose (see Table 3.2). The values of $\ln[T(x) - T(\infty)]$ are plotted against dose (x) in Fig. 3.6.

The constants A, B, and C corresponding to the observations of Example 9 are

$$A = 59.5, \qquad B = -1.285, \qquad C = 40.5$$

Of these, A and B were obtained from Fig. 3.6 as follows. Intercept of the

Table 3.2

Linearization of the Data for the Pyridoxine Dose–Response
Relationship (Example 9)

Dose (x)	$T(x)$	$T(x) - T(\infty)$	$\ln[T(x) - T(\infty)]$
0.00	100.0	59.5	4.086
0.25	81.5	41.0	3.714
0.50	69.0	28.5	3.499
0.75	62.0	21.5	3.068
1.00	57.0	16.5	2.803
1.25	52.5	12.0	2.485
1.50	49.5	9.0	2.197
1.75	47.0	6.5	1.872
2.00	45.0	4.5	1.504

response line on the vertical axis at zero dose is 4.09. Using tables of natural
logarithms,

$$A = \text{antilog } 4.09 = 59.5$$

The slope B is obtained from the graph over the entire dose range. The
function of response corresponding to $x = 0$ is 4.09, and that corresponding
to $x = 2$ (coded dose 8) is 1.52, so that

$$B = (1.52 - 4.09)/2 = -1.285$$

The value of C was obtained in Example 9 as

$$C = T(\infty) = 40.5$$

The applicability of Eq. (3.1) may be verified by substituting these values
for A, B, and C as well as any chosen value for x.
Thus Eq. (3.1) becomes

$$T(x) = 59.5e^{-1.285x} + 40.5$$

Putting $x = 1.25$,

$$T(x) = 59.5e^{-(1.285 \times 1.25)} + 40.5$$

Rearranging and taking natural logarithms,

$$\ln[T(x) - 40.5] = \ln 59.5 - (1.285 \times 1.25) = 4.086 - 1.606 = 2.480$$

Taking antilogs,

$$T(x) - 40.5 = 11.9$$

so that

$$T(x) = 11.9 + 40.5 = 52.4$$

This is very close to the actually observed value, 52.5.

In routine work it is not necessary to evaluate A and B, and so it is convenient to plot graphs using logarithms to base 10.

The relationship between Eq. (3.3) and the ideal case may be seen by putting $T(\infty) = 0$, which represents infinite response to infinite dose. Expressed more realistically this means that there are no limiting factors other than the g.p.s. that is to be determined. In this case a plot of log $T(x)$ versus dose would be a straight line, as also would a plot of $2 - \log T(x)$ (i.e., absorbance) versus dose. The latter is a form commonly used in turbidimetric assays.

It will be seen by comparing Figs. 3.5 and 3.6 how larger values of $T(\infty)$ correspond to increased curvature in a plot of absorbance versus dose.

Values of $T(\infty)$ for particular assays are stated to be reasonably constant provided that assay variables are strictly controlled. Nevertheless Tsuji calculates a value daily from the evidence of the day's assay.

It should be noted that this procedure was designed for use with a computer. The calculation of $T(\infty)$ is a simple matter, however, and so the method may be applied to manual calculations. The straight line responses obtained pass through a common point at zero dose, and so potencies can be calculated by the slope ratio method (see Section 3.4)

3.4   Slope Ratio Assays—An Unbalanced Design

When the ideal conditions have been met, responses in vitamin and amino acid assays increase directly as the dose. Consequently one dose–response line can be made to correspond to another by changing the dose scale by a constant multiplier. This means that potency estimated for a sample is independent of the concentration at which it is assayed. A requirement for validity is intersection of standard and sample dose–response lines at zero dose. Potency ratio of sample to standard test solutions is estimated from the equation

$$R = b_T/b_S \tag{3.5}$$

where b_T is slope of the sample line and b_S is the slope of the standard line. Such assays are known as "slope ratio assays."

Examples may now be given of the graphical and arithmetical calculation procedures. The response lines shown in Fig. 3.7 for Example 10, the assay

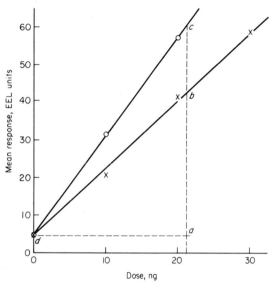

Fig. 3.7. Responses lines for an unbalanced slope ratio assay. The data refer to the assay of pantothenic acid using *Lactobacillus arabinosus* (Example 10). ×, standard response line; ○, sample response line.

of pantothenic acid, are nearly ideally straight. In this assay, the two lines intersect at zero dose but not at zero response. The latter is rare because carry over of growth-promoting substance in the inoculum permits a small amount of growth.

Since in this case the lines are straight, procedures based on ratio of the slopes can be applied.

Example 10: Assay of Pantothenic Acid

Test organism: *Lactobacillus arabinosus* ATCC 8034

Observations: Arbitrary units measured by EEL nephelometer are given in Table 3.3.

Reference standard: A solution containing 1 mg/ml of calcium pantothenate was diluted to levels of 10, 20, and 30 ng/ml

Sample preparation: Multivitamin tablets containing nominally 10 mg/tablet. Weight of 10 tablets = 3.04 g. The ten tablets were powdered and 48.7 mg of the powder was diluted to 1 liter. Further dilutions from this solution were made:

$$5 \text{ ml} \rightarrow 500 \text{ ml} \quad \text{and} \quad 5 \text{ ml} \rightarrow 250 \text{ ml}$$

giving assumed concentrations of about 16 and 32 ng/ml, respectively. However, for purposes of calculating estimated potency from the observed responses, it is convenient to regard these solutions as having nominal potencies of 10 and 20 ng/ml and then to compare them with the two lower levels of standard test solution.

One milliliter of each standard test solution was added to each of four tubes, 1 ml of each sample test solution to each of three tubes, and 1 ml of water to each of two tubes to represent the zero control. Then 9 ml of broth was added to each tube. Each tube was inoculated with one drop of a suspension of the test organism then incubated for 18 hours at 37°C.

Observations: See Table 3.3.

Table 3.3

Dose (ng/tube)		Response (EEL units)				Total response	Mean response
Standard	10	20	22	19	22	83	20.75
(actual dose)	20	41	42	40	43	166	41.50
	30	59	58	57	61	235	58.75
Sample	10	30	32	33		95	31.67
(nominal dose)	20	56	57	59		172	57.33
Zero control	0	4	5			9	4.50

The best estimate of the slope ratio obtainable from the graph of Fig. 3.7 is by the intercepts b and c of the vertical dashed line ac on the two response lines and the horizontal dashed line ad, which passes through the point of intersection of the two response lines with the vertical axis. The slope ratio and therefore the estimate of potency ratio are obtained as

$$(c - a)/(b - a) = (60.0 - 4.5)/(42.2 - 4.5) = 1.472$$

Potency of each tablet of average weight 304 mg (calculating from lowest dilutions) is

$$\frac{10 \times 1.472 \times 500 \times 1000 \times 304}{5 \times 48.7 \times 10^6} = 9.2 \quad \text{mg/tablet}$$

A result obtained graphically in this way is adequate for many purposes and is obtained less laboriously than by the arithmetical method that will be illustrated shortly.

However, this design of assay is not to be recommended, as much more practical effort has been devoted to measuring the standard response (twelve observations) than to measuring the sample response (six observations). A balanced design, that is, one that uses the same number of responses for sample and standard at the same nominal dose levels, is not only more efficient, it is also amenable to much simpler arithmetical treatment. The calculation of results from balanced assays is shown in Section 3.8.

3.5   Criteria of Validity

An essential requirement for slope ratio assays is that the dose–response lines should intersect at zero dose. Failure of the sample and standard dose–response lines to intersect at zero dose may indicate an invalid assay. In practice some deviation is to be expected due to random variations. Nevertheless, when deviation is slight, estimation of potency should be based on graphs in which the response lines are drawn to intersect at zero dose. Intersecting of the lines at zero response is rare and should not be expected. Graphs resembling those in Fig. 3.8 may be obtained. Type (a) response appears to indicate that the inoculum, which has been grown in a complete medium, has carried over a small reserve of the growth factor that is to be assayed. Type (b) response may indicate an apparent threshold quantity of growth factor that is required before growth can begin. This quantity would be indicated by the intercept of the standard response line on the dose scale at zero response.

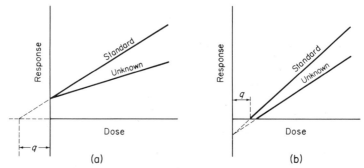

Fig. 3.8. An illustration of variation in the form of response curves in slope ratio assays. Ideally, standard and sample response lines should intersect at zero response to zero dose. In graph (a) it appears that a small constant quantity of the substance to be estimated is present in all tubes. This quantity is indicated by the distance q on the extrapolated dose scale. In graph (b) there is an apparent threshold quantity requirement of the substance to be estimated. This quantity is indicated by the distance q on the dose scale.

There are several possible explanations for this apparent threshold quantity. Adsorption on the glass might account for some loss of the essential g.p.s. According to the laws of adsorption this would not be a constant amount but would increase with increasing concentration in the liquid phase. However, the precision of the assay may be inadequate to detect such changes, so that the adsorbed amount may seem to be roughly constant.

Alternatively, a portion of the essential g.p.s. might be bound to some component of the medium. In the case of vitamin B_{12} assays, it has been shown (Skeggs, 1963) that a deficiency in reducing agent in the medium causes this form of curve. Addition of the threshold quantity of growth factor to the test media may raise the origin of the graph. However, such deviations, if they do occur and are reasonably small, do not present any problems in obtaining a result. Larger deviations should be investigated to find the practical causes.

In case (b), as the observations corresponding to zero dose must be either zero or positive, they would not lie on the lines indicated by the sample and standard responses. It is therefore necessary to discard these observations.

3.6   Multiple Linear Regression Equations

It is normally adequate in routine work to obtain a potency estimate from an unbalanced assay graphically. However, if an estimate of the reliability of the result is also required, the result must be obtained arithmetically, and the calculation then continued to give confidence limits as described in Chapter 7.

It should be noted that this tedious calculation procedure is necessitated by the unbalanced design. Contrast this calculation with the simple procedure for balanced assay designs such as shown in Sections 3.8 and 3.10. Simple linear regression equations were introduced in Section 2.9. Although it is possible to calculate two separate regression equations for standard and sample response lines, the intersection point of the two lines will probably fail to coincide exactly at zero dose.

The lines of best fit must conform to the requirement for a valid assay that they should intersect at zero dose. These lines of best fit are obtained by multiple linear regression equations. The principle of these equations is shown as follows:

(1)   In a simple linear dose–response graph the relationship between the deviation of any hypothetical dose x from the mean dose $\bar{x}$ and the deviation of the expected corresponding response y_e from the mean response $\bar{y}$ may be expressed as

$$(x - \bar{x})b = (y_e - \bar{y}) \tag{3.6}$$

The slope b is determined from the simple regression Eq. (2.6) rewritten as

$$S(x - \bar{x})^2 b = S(y - \bar{y})(x - \bar{x})$$

which is related to Eq. (3.6) in that (a) y_e is replaced by an actual observation, (b) both sides are multiplied by $(x - \bar{x})$, and (c) the summation for all doses and responses is obtained.

(2)   In the case of two response lines corresponding to standard and sample, all responses are considered together; thus $\bar{y}$ respresents the mean of all responses, standard and sample.

Individual deviations from $\bar{y}$ are related to deviations from both the mean standard dose $\bar{x}_S$ and mean sample dose $\bar{x}_T$. For example, if the mean dose of standard $\bar{x}_S$ for all tubes in the assay is 0.08 μg and similarly the mean nominal dose for sample $\bar{x}_T$ is 0.06 μg, then a standard test solution containing 0.10 μg of standard (and 0.00 μg of sample) deviates from the mean standard dose thus:

$$(x_S - \bar{x}_S) = (0.10 - 0.08) = +0.02 \quad \mu g$$

and similarly from the mean sample dose:

$$(x_T - \bar{x}_T) = (0.00 - 0.06) = -0.06 \quad \mu g$$

Thus for any dose of sample or standard, the deviation of the expected response y_e from $\bar{y}$ (the mean of all standard and sample responses) may be expressed as the sum of two components:

(a)   The response deviation $(x_S - \bar{x}_S)b_S$ from overall mean response $\bar{y}$ due to standard component of dose deviation from $\bar{x}_S$.

(b)   The response deviation $(x_T - \bar{x}_T)b_T$ from $\bar{y}$ due to sample component of dose deviation from $\bar{x}_T$. (The "standard component" of a sample test solution is of course zero and vice versa, thus giving negative components to the net deviation.) Thus we may write for overall response deviation

$$(x_S - \bar{x}_S)b_S + (x_T - \bar{x}_T)b_T = (y_e - \bar{y}) \tag{3.7}$$

where b_S and b_T represent slopes or regression coefficients for standard and sample, respectively.

By analogy with the simple regression equation rewritten as in (1) above, two equations may be derived from Eq. (3.7) replacing y_e by y and then multiplying first by $(x_S - \bar{x}_S)$ to obtain the first of these equations and then by $(x_T - \bar{x}_T)$ to obtain the second; thus,

$$S(x_S - \bar{x}_S)^2 b_S + S(x_S - \bar{x}_S)(x_T - \bar{x}_T)b_T = S(x_S - \bar{x}_S)(y - \bar{y}) \tag{3.8a}$$

$$S(x_T - \bar{x}_T)(x_S - \bar{x}_S)b_S + S(x_T - \bar{x}_T)^2 b_T = S(x_T - \bar{x}_T)(y - \bar{y}) \tag{3.8b}$$

These are more conveniently written using abbreviated symbols analogous to those defined in expressions (2.7) and (2.8):

$$b_S S_{x_S x_S} + b_T S_{x_S x_T} = S_{x_S y} \tag{3.9a}$$

$$b_S S_{x_S x_T} + b_T S_{x_T x_T} = S_{x_T y} \tag{3.9b}$$

The values of $S_{x_S x_S}$, $S_{x_T x_T}$, $S_{x_S x_T}$, $S_{x_S y}$, and $S_{x_T y}$ are calculated from the observed results; then substituting their values in the two equations, the values of b_S and b_T may be found.

3.7 Potency Computation from an Unbalanced Design

This calculation is now applied to the observations of Example 10. In order to avoid the possibility of the nonmathematical reader being confused by a mass of apparently meaningless algebraic expressions, this first example of a multiple regression equation is worked out in full without any of the normal convenient shortcuts. For convenience of calculation doses are coded as 1, 2, and 3 for standard and 1 and 2 for sample.

To obtain the values of $S_{x_S x_S}$, $S_{x_T x_T}$, and $S_{x_S x_T}$:

(1) Calculate $\bar{x}_S$ and $\bar{x}_T$:

$$Sx_S = (4 \times 1) + (4 \times 2) + (4 \times 3) = 24$$

As there are in all 20 responses in the test, the mean dose of standard in *all tubes* is

$$\bar{x}_S = 24/20 = 1.20$$

Similarly

$$Sx_T = (3 \times 1) + (3 \times 2) = 9$$

and

$$\bar{x}_T = 9/20 = 0.45$$

(2) Tabulate the values of the deviations $x_S - \bar{x}_S$ and $x_T - \bar{x}_T$ (Table 3.4).

(3) Calculate $S_{x_S x_S}$, $S_{x_T x_T}$, and $S_{x_S x_T}$ from these deviations and the corresponding replications, which are standard $= 4$, sample $= 3$, zero control $= 2$.

Table 3.4

Deviations of individual coded doses x_S and x_T from their respective mean coded dose levels $\bar{x}_S$ and $\bar{x}_T$ throughout all tubes in the assay[a]

Preparation	x_S	x_T	$x_S - \bar{x}_S$	$x_T - \bar{x}_T$
Standard	1	0	-0.20	-0.45
	2	0	$+0.80$	-0.45
	3	0	$+1.80$	-0.45
Sample	0	1	-1.20	$+0.55$
	0	2	-1.20	$+1.55$
Zero control	0	0	-1.20	-0.45

[a] $\bar{x}_S = 1.20$, $\bar{x}_T = 0.45$.

Thus:

$$S_{x_Sx_S} = S(x - \bar{x}_S)^2 = 4[(-0.20)^2 + (0.80)^2 + (1.80)^2]$$
$$+ 3[(-1.20)^2 + (-1.20)^2] + 2[(-1.20)^2] = 27.20$$

$$S_{x_Tx_T} = S(x - \bar{x}_T)^2 = 4[(-0.45)^2 + (-0.45)^2 + (-0.45)^2]$$
$$+ 3[(0.55)^2 + (1.55)^2] + 2[(-0.45)^2] = 10.950$$

$$S_{x_Sx_T} = S(x_S - \bar{x}_S)(x_T - \bar{x}_T)$$
$$= 4[(-0.20 \times -0.45) + (0.80 \times -0.45) + (1.80 \times -0.45)]$$
$$+ 3[(-1.20 \times 0.55) + (-1.20 \times 1.55)] + 2[(-1.20 \times -0.45)]$$
$$= -10.800$$

To obtain the values of S_{xsy}:

(1) Calculate $\bar{y}$. The total of all responses $S_y = 760$ and as there are in all 20 responses, the mean response is

$$\bar{y} = 760/20 = 38$$

(2) Tabulate the values of the deviations $y - \bar{y}$ (Table 3.5).

Table 3.5

The experimental data of Example 10[a]

	1st tube		2nd tube		3rd tube		4th tube	
	y	$(y - \bar{y})$	y	$(y - \bar{y})$	y	$(y - \bar{y})$	y	$(y - \bar{y})$
Standard	20	−18	22	−16	19	−19	22	−16
	41	+3	42	+4	40	+2	43	+5
	59	+21	58	+20	57	+19	61	+23
Sample	30	−8	32	−6	33	−5	—	—
	56	+18	57	+19	59	+21	—	—
Zero control	4	−34	5	−33	—	—	—	—

[a] Responses y are expressed as deviations $y - \bar{y}$, from the mean of responses $\bar{y}$ for all tubes in the assay.

(3) Calculate S_{xsy} and S_{xTy} by multiplying these response deviations by the corresponding dose deviations and summing appropriately; thus:

$$S_{xsy} = S(x_S - \bar{x}_S)(y - \bar{y})$$
$$= -0.20(-18 - 16 - 19 - 16) + 0.80(3 + 4 + 2 + 5)$$
$$+ 1.80(21 + 20 + 19 + 23) - 1.20(-8 - 6 - 5)$$
$$- 1.20(18 + 19 + 21) - 1.20(-34 - 33) = 208.00$$

$$S_{x_Ty} = S(x_T - \bar{x}_T)(y - \bar{y})$$
$$= -0.45(-69) - 0.45(14) - 0.45(83) + 0.55(-19)$$
$$+ 1.55(58) - 0.45(-67) = 97$$

The values are summarized for convenience:

$$S_{x_Sx_S} = 27.20, \qquad S_{x_Tx_T} = 10.95, \qquad S_{x_Sx_T} = -10.80$$
$$S_{x_Sy} = 208.00, \qquad S_{x_Ty} = 97.00$$

The regression coefficients b_S and b_T may be obtained by substituting the above values into Eqs. (3.9a) and (3.9b):

(i) $27.20b_S - 10.80b_T = 208$

(ii) $-10.80b_S + 10.95b_T = 97$

Solve these equations by first multiplying (i) by $1095/1080 = 1.013888$ giving

(ia) $27.58b_S - 10.95b_T = 210.89$

(ii) $-10.80b_S + 10.95b_T = 97.00$

$\overline{16.78b_S = 307.89}$

Thus

$$b_S = 307.89/16.78 = 18.349$$

Second, multiply (ii) by $272/108 = 2.518518$:

(i) $27.20b_S - 10.80b_T = 208.00$

(iia) $-27.20b_S + 27.58b_T = 244.30$

$\overline{16.78b_T = 452.30}$

Thus

$$b_T = 452.3/16.78 = 26.955$$

By Eq. (3.5),

$$R = 26.955/18.349 = 1.469$$

However, it is preferable to obtain b_S and b_T directly by substituting in the matrix form of the equation; thus,

$$b_S = \frac{(S_{x_Sy})(S_{x_Tx_T}) - (S_{x_Ty})(S_{x_Sx_T})}{(S_{x_Sx_S})(S_{x_Tx_T}) - (S_{x_Sx_T})^2} \tag{3.10a}$$

$$b_T = \frac{(S_{x_Ty})(S_{x_Sx_S}) - (S_{x_Sy})(S_{x_Sx_T})}{(S_{x_Sx_S})(S_{x_Tx_T}) - (S_{x_Sx_T})^2} \tag{3.10b}$$

giving

$$b_S = \frac{(208.00)(10.95) - (97.00)(-10.80)}{(27.20)(10.95) - (-10.80)^2} = \frac{3325.20}{181.20} = 18.351$$

$$b_T = \frac{(97.00)(27.20) - (208.00)(-10.80)}{(27.00)(10.95) - (-10.80)^2} = \frac{4884.80}{181.20} = 26.958$$

By Eq. (3.5),

$$R = 26.958/18.351 = 1.469$$

which corresponds to a sample potency of 9.15 mg of calcium pantothenate per tablet of average weight. This should be compared with the value of 9.2 mg/tablet obtained graphically in Section 3.4.

3.8   Balanced Slope Ratio Assays with Linearity Check

As described in Section 3.5, responses to zero dose sometimes fail to lie on the lines corresponding to standard and sample responses. Thus if it is required to check that response lines are straight, a minimum of three dose levels is necessary.

A balanced slope ratio assay incorporating three dose levels for standard and sample is illustrated in another assay of pantothenic acid in Example 11.

Example 11:   Assay of Pantothenic Acid

Test organism: *Lactobacillus arabinosus* ATCC 8034
Observations: Arbitrary units measured by EEL nephelometer
Preparation of test solutions
Standard: A solution containing 1 mg/ml of calcium pantothenate was diluted to levels of 10, 20, and 30 ng/ml.
Sample preparation: Multivitamin tablets containing nominally 5 mg of calcium pantothenate per tablet. Weight of 10 tablets = 4.29 gm. The ten tablets were powdered and 51.9 mg of the powder was diluted to 1 liter. Further dilutions were made from this solution thus:

$$5 \text{ ml} \rightarrow 250 \text{ ml}$$
$$10 \text{ ml} \rightarrow 250 \text{ ml}$$
$$15 \text{ ml} \rightarrow 250 \text{ ml}$$

giving nominal concentrations of 10, 20, and 30 ng/ml of calcium pantothenate in the test solution. A set of four tubes was prepared for each treatment and 1 ml of test solution was pipetted into each tube. Similarly 1 ml of water was pipetted into each of a set of four tubes for the zero control.

Nine milliliters of assay medium was added to each of the 28 tubes. Each tube was inoculated with one drop of a suspension of the test organism then all were incubated for 18 hours at 37°C.

Observations: See Table 3.6.

Table 3.6

Dose ng/tube		Response (EEL units)				Total response	Mean response
Standard	10	22	22	19	20	83	20.75
	20	39	41	38	40	158	39.50
	30	57	58	55	54	224	56.00
Sample	10	30	26	29	26	111	27.75
	20	51	51	47	49	198	49.50
	30	76	72	71	75	294	73.50
Zero control	0	3	4	3	4	14	3.50

Mean responses are plotted against dose in Fig. 3.9, and the sample potency ratio may be obtained from this graph as

$$(c - a)/(b - a) = (68.0 - 3.5)/(52.5 - 3.5) = 1.316$$

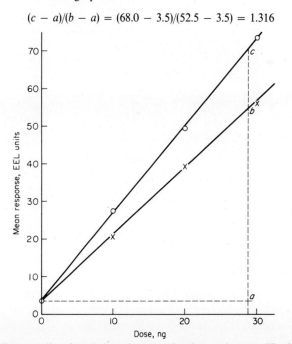

Fig. 3.9. Response lines for a balanced seven point slope ratio assay. The data refer to the assay of pantothenic acid using *Lactobacillus arabinosus* (Example 11). ×, standard; ○, sample.

Sample potency is obtained relative to low dose standard (10 ng) thus:

$$(10 \times 1.316 \times 250 \times 1000 \times 429)/(5 \times 51.9 \times 1000 \times 1000) = 5.44 \text{ mg/tablet}$$

As in Example 10, the result may be obtained arithmetically. The calculation is slightly simplified due to this being a balanced design. Using coded doses 1, 2, and 3 and applying the multiple regression equation but using this time the normal short form of the equation, first calculate

$$Sx_S = 24, \qquad Sx_T = 24, \qquad n = 28, \qquad Sy = 1082$$

$$S_{x_Sx_S} = S(x_S)^2 - (Sx_S)^2/n = [4(1^2 + 2^2 + 3^2)] - [(24)^2/28] = 35.429$$

Similarly

$$S_{x_Tx_T} = 35.429, \qquad S_{x_Sx_T} = [S(x_S)(x_T)] - [(Sx_S)(Sx_T)/n]$$

Whenever x_S has a value other than zero, x_T is zero, and vice versa; thus the first term of this expression is always zero, giving

$$S_{x_Sx_T} = 0 - (24 \times 24)/28 = -20.571$$

$$\begin{aligned} S_{x_Sy} &= [S(x_Sy)] - [(Sx_S)(Sy)/n] \\ &= [(1 \times 83) + (2 \times 158) + (3 \times 224)] - [(24)(1082)/28] \\ &= 1071 - 927.428 = 143.572 \end{aligned}$$

$$\begin{aligned} S_{x_Ty} &= [S(x_Ty)] - [(Sx_T)(Sy)/n] \\ &= [(1 \times 111) + (2 \times 198) + (3 \times 294)] - [(24)(1082)/28] \\ &= 1389 - 927.428 = 461.572 \end{aligned}$$

Substituting these values in the multiple regression Eqs. (3.9a) and (3.9b), the following are obtained:

$$35.429b_S - 20.571b_T = 143.572$$

$$-20.571b_S + 35.429b_T = 461.572$$

On solving these simultaneous equations it is found that

$$b_S = 17.5248, \qquad b_T = 23.2033$$

By Eq. (3.5),

$$R = 23.2033/17.5248 = 1.324$$

leading to a potency of 5.47 mg/tablet.

This figure for potency ratio differs only slightly from the value 1.316 that was obtained graphically with relatively little effort and so the reader may wonder whether the calculation by multiple regression equation is worthwhile.

It should be borne in mind, however, that the labor of computation may be much reduced in routine work by using a standard pattern of assay. If the replication of standard, samples, and blanks and the number of dose levels do not change from assay to assay, then the values for n, $S_{x_S x_S}$, $S_{x_T x_T}$, and $S_{x_S x_T}$ are constant and $S_{x_S y}$ and $S_{x_T y}$ may be calculated simply.

Thus, in Example 11.

$$S_{x_S y} = S(x_S y) - 0.8571(Sy) = 1071 - 0.8571(1082) = 143.618$$

Similarly,

$$S_{x_T y} = S(x_T y) - 0.8571(Sy) = 1389 - 0.8571(1082) = 461.618$$

The values for $S_{x_S y}$ and $S_{x_T y}$ are then substituted in the partially worked out matrix equations to obtain b_S and b_T directly:

$$b_S = [35.429(S_{x_S y})] + [20.571(S_{x_T y})/832.048]$$
$$= [35.429(143.618)] + [20.571(461.618)/832.048] = 17.5280$$

Similarly,

$$b_T = [35.429(461.618)] + [20.571(143.618)/832.048] = 23.2066$$

By Eq. (3.5),

$$R = 23.2066/17.5280 = 1.324$$

which is exactly the same figure as obtained by the previous calculation. The work may be eased still further by the use of proformas in which all constant figures are printed.

3.9   Simplified Computation of Potency Ratio from Balanced Assays

As has been demonstrated, results of both balanced and unbalanced slope ratio assays may be obtained by fitting a multiple regression equation to the observations. However, the calculations are rather lengthy for nonroutine work and would become very tedious if applied to multiple assays that include a comparison of more than one sample with the standard.

The results of balanced assays may be obtained more easily by substituting in a simple formula appropriate to the design of the assay. Formulas for multiple assays may be derived. The principles involved are exemplified by the derivation of a formula for a balanced seven point assay such as described in Example 11.

The ideal response is two straight lines for sample and standard intersecting on the y axis at a short distance a' from zero. The two lines may be represented by the equations

$$y_S = a' + b_S x \qquad (3.11a)$$

$$y_T = a' + b_T x \qquad (3.11b)$$

for standard and unknown, respectively, where b_S is the increase in response for unit dose of standard, i.e., it is the slope for standard, and b_T is the corresponding increase for unit dose of the unknown. Both b_S and b_T refer to increases in a single tube. The ideal responses (y) to the seven doses may be expressed as shown in Table 3.7.

Table 3.7

Preparation	Dose x	Response y
Zero control	0	a'
Standard	1	$a' + b_S$
	2	$a' + 2b_S$
	3	$a' + 3b_S$
Unknown	1	$a' + b_T$
	2	$a' + 2b_T$
	3	$a' + 3b_T$

As was explained in Chapter 2, a more precise estimate of slope may be obtained from a wide difference in dose than from a small difference.

Provided that instrument error is constant over the range of responses used (a reasonable assumption), allowance for the more precise estimate can be made by weighting the response according to the dose interval with respect to zero. Thus for the standard, a weighted sum of the responses is expressed by the symbol T_S which is equal to

$$(0 \times \text{response to zero dose}) + (1 \times \text{response to dose 1})$$
$$+ (2 \times \text{response to dose 2}) + \ldots, \text{etc.}$$

or

$$T_S = 0(a') + 1(a' + b_S) + 2(a' + 2b_S) + 3(a' + 3b_S) = 6a' + 14b_S \quad (3.12)$$

Similarly, for the unknown,

$$T_T = 6a' + 14b_T \quad (3.13)$$

Combining T_S and T_T as $S(T_i)$, where T_i represents the weighted sum of responses to any treatment,

$$S(T_i) = 12a' + 14b_S + 14b_T \quad (3.14)$$

The unweighted sum of all responses $S(y)$ is given by the expression

$$S(y) = a' + (a' + b_S) + (a' + 2b_S) + (a' + 3b_S)$$
$$+ (a' + b_T) + (a' + 2b_T) + (a' + 3b_T)$$

which simplifies to

$$S(y) = 7a' + 6b_S + 6b_T \quad (3.15)$$

Then an expression for a' may be obtained in terms of $S(y)$ and $S(T_i)$ by first multiplying Eq. (3.15) by 7, then subtracting from it Eq. (3.14) multiplied by 3; thus

$$7S(y) = 49a' + 42b_S + 42b_T$$
$$3S(T_i) = 36a' + 42b_S + 42b_T$$

so that

$$7S(y) - 3S(T_i) = 49a' - 36a' = 13a'$$

and

$$a' = \frac{[7S(y) - 3S(T_i)]}{13} \tag{3.16}$$

Thus a has been expressed in a simple form incorporating the contributions from all observations.

The values of the slopes b_i may then be readily obtained by rewriting Eqs. (3.12) and (3.13) as

$$b_S = \frac{(T_S - 6a')}{14} \quad \text{and} \quad b_T = \frac{(T_T - 6a')}{14}$$

These formulas for a', b_S, and b_T are special cases of the general formula derived by Bliss (1952), who gives

$$a' = \frac{2(2k + 1)S(y) - 6S(T_i)}{N(k - 1) + 3h'(k + 1)} \tag{3.17}$$

$$b_i = \frac{3}{2k + 1}\left[\frac{2T_i}{fk(k + 1)} - a'\right] \tag{3.18}$$

where f is the degree of replication at each dose level (standard, sample), k the number of dose levels, h' the replication at zero dose, and N the total number of tubes. To show the identity with the specific example previously derived, substitute in Bliss' formulas (3.17) and (3.18) the values $f = 1$, $k = 3$, $h' = 1$, and $N = 7$. Thus

$$a' = \frac{2[(2 \times 3) + 1]S(y) - 6S(T_i)}{7(3 - 1) + (3 \times 1)(3 + 1)} = \frac{14S(y) - 6S(T_i)}{14 + 12}$$

which is identical with the original result for a'. Also

$$b_S = \frac{3}{(2 \times 3) + 1}\left[\frac{2T_i}{(1 \times 3)(3 + 1)} - a'\right] = \frac{3}{7}\left[\frac{2T_S}{12} - a'\right] = \frac{T_S - 6a'}{14}$$

which is identical with the original value for b_S.

The Bliss formulas may be applied to both simple and multiple assays. When responses to zero dose are not collinear with other responses, then N and $S(y)$ are replaced by N^* and $S(y)^*$, respectively, which exclude the zero dose tubes.

The virtue of mathematical formulas is that they remove the error of visual assessment of the best straight lines. They are therefore capable of yielding the most accurate estimate of potency that the observations permit.

If a graph is drawn, however, it is immediately obvious if the observed responses fail to conform to the assumed mathematical relationship, i.e., straight lines intersecting at zero dose. The ideal routine then is to:

(1)   Either draw a graph or inspect the observations to see if they appear to conform to the assumed mathematical relationship (the latter is very simple and rapid).

(2)   If the correct relationship is confirmed, then proceed with the calculation.

The observations of Example 11 may now be processed by these formulas. The essential data are

$$S(y) = 1082$$
$$T_S = 83 + (2 \times 158) + (3 \times 224) = 1071$$
$$T_T = 111 + (2 \times 198) + (3 \times 294) = 1389$$
$$S(T_i) = T_S + T_T = 2460$$

Substitution in Bliss' general formulas of the constants for this assay design, $f = 4, k = 3, h' = 4, N = 28$, lead by Eq. (3.17) to

$$a' = \frac{2[(2 \times 3) + 1]S(y) - 6S(T_i)}{28(3 - 1) + (3 \times 4)(3 + 1)} = \frac{14S(y) - 6S(T_i)}{104}$$

and by Eq. (3.18) to

$$b_i = \frac{3}{(2 \times 3) + 1}\left[\frac{2T_i}{(4 \times 3)(3 + 1)} - a'\right] = \frac{T_i - 24a'}{56}$$

Substitution of the values for $S(y)$, $S(T_i)$, and T_i leads to

$$a' = [(14 \times 1082) - (6 \times 2460)]/104 = 3.73$$
$$b_S = [1071 - (24 \times 3.73)]/56 = 17.53$$
$$b_T = [1389 - (24 \times 3.73)]/56 = 23.21$$

By Eq. (3.5),

$$R = 23.21/17.53 = 1.324$$

which again leads to the same estimate of potency of 5.47 mg/tablet.

3.10   Multiple Assays by the Slope Ratio Method

When an assay method is in regular use and it is known that the response lines are straight, the analyst will naturally wish to employ a design that will give potency estimate of the required precision with the minimum effort. It follows that superfluous effort should not be expended on checks for curvature.

In a critical review of assay design, Wood (1946) showed that the mathematically ideal design for the comparison of the preparations by the slope ratio assay comprised one zero dose control and one unit dose level of each preparation. However, such a design provided no check that the zero control response actually lay on the dose–response line and so was impracticable. Wood therefore recommended a compromise design including dose levels in the ratio 1:2 of each preparation plus the zero dose control. This has become widely known as the five point common zero assay. Calculation of results by either multiple regression equations or by the Bliss formula is a simple matter.

In routine work it is usual to compare several samples with the standard in a single test. As the number of samples increases, the solution of the multiple regression equations becomes more laborious. It is therefore highly desirable to use completely balanced designs in order that results may be calculated simply by the formulas of Bliss [Eqs. (3.17) and (3.18)].

The simultaneous assay of three samples using only two dose levels is illustrated in Example 12 on the assay of folic acid.

Example 12:   Assay of Folic Acid

Test organism: *Streptococcus faecalis* ATCC 10541
Observations: Arbitrary units measured by EEL nephelometer
Preparation of test solutions:
Standard: A solution containing 1 mg/ml of folic acid was diluted to levels of 1 and 2 ng/ml.
Sample weighings and dilutions: Sample A, multivitamin tablets, nominal folic acid content 100 μg/tablet, weight of ten tablets is 4.427 g. Ten tablets were reduced to powder and a weight equivalent to about one tablet was triturated with water to prepare nominal dilutions of 1 and 2 ng/ml. The dilution scheme to 1 ng/ml follows:

0.4378 g → 500 ml : 5 ml → 50 ml : 5 ml → 100 ml : 5 ml → 50 ml

Sample B, multivitamin tablets, nominal folic acid content 100 μg/tablet, weight of ten tablets is 4.518 g. Preparation of test solutions as for Sample A using 0.4223 g of powdered tablets.

Sample C, tablets of folic acid U.S.P., nominal strength 400 μg/tablet, weight of 10 tablets is 1.032 g. The dilution scheme to the low dose test solution follows:

0.0983 g → 200 ml : 5 ml → 50 ml : 5 ml → 50 ml : 5 ml → 100 ml

Tubes were set up in quadruplicate with 1 ml of test solution (1 ml of water in the case of the zero control) and 9 ml of the assay medium. Tubes were each inoculated with one drop of a suspension of the test organism and incubated for 18 hours at 37°C.

Observations are recorded in Table 3.8.

Table 3.8

Data from the Assay of Folic Acid (Example 12)

Preparation	Dose (x) ng	Response (y) EEL units				Response totals	Mean response
Standard	1	33	31	29	29	122	30.50
	2	70	68	71	67	276	69.00
Sample A	1	34	31	35	33	133	33.25
	2	75	76	79	76	306	76.50
Sample B	1	31	29	31	27	118	29.50
	2	67	67	64	66	264	66.00
Sample C	1	30	33	31	31	125	31.25
	2	72	70	74	74	290	72.50
Zero control	0	5	4	4	6	19	4.75

Mean responses are plotted in Fig. 3.10, from which it is seen that the responses to zero dose do not lie on the response lines for sample and standard. Therefore, in using Bliss' general formula for the calculation of potency ratios it is necessary to choose the form that omits the responses to zero dose. While it is always instructive to draw a rough graph before applying the purely arithmetical calculation, inspection of the tabulated mean responses would also reveal that zero dose should be omitted.

Potency ratios may be obtained graphically by

(i)   drawing a horizontal line through the point at which the four response lines intersect at zero dose,

(ii)   measuring the distances between the intercepts on this horizontal line and the individual response lines made by a convenient vertical line.

Potencies of test solutions are in the same ratio as these distances.

From Fig. 3.10 the horizontal line crosses the vertical axis at -8.5. A vertical line drawn at the point corresponding to a dose of 1.8 ng gives intercepts on the response lines of 67.5, 58.5, 64.0, and 61.5 for Samples A, B, C, and standard, respectively.

This leads to potency ratios of

$$\text{Sample A:} \quad [67.5 - (-8.5)]/[61.5 - (-8.5)] = 1.085$$
$$\text{Sample B:} \quad [58.5 - (-8.5)]/[61.5 - (-8.5)] = 0.957$$
$$\text{Sample C:} \quad [64.0 - (-8.5)]/[61.5 - (-8.5)] = 1.036$$

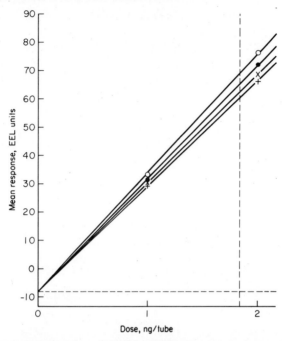

Fig. 3.10. Response lines for a balanced multiple two dose level slope ratio assay. The data refer to the assay of folio acid using *Streptococcus faecalis* (Example 12). ×, standard; ◯, Sample A; +, Sample B; ●, Sample C.

From these ratios and dilutions to the low dose level sample potencies are

Sample A: $\dfrac{1.085 \times 100 \times 50 \times 500 \times 4.427 \times 1}{5 \times 5 \times 0.4378 \times 10} = 110$ μg/tablet

Sample B: $\dfrac{0.957 \times 100 \times 50 \times 500 \times 4.518 \times 1}{5 \times 5 \times 0.4223 \times 10} = 102$ μg/tablet

Sample C: $\dfrac{1.036 \times 100 \times 50 \times 50 \times 200 \times 1.032 \times 1}{5 \times 5 \times 5 \times 0.0983 \times 10} = 435$ μg/tablet

To illustrate nongraphical computation procedures, the potency of Sample A is first calculated as for an assay of a single sample.

Inspection of observations reveals that the response to zero dose is not collinear with either sample or standard response lines. The latter would both cut the zero axis at about -8 to -10. This is close enough to assume a common zero (i.e., there is nothing to suggest that the assay is invalid), and so potency ratio may be calculated using Bliss' formulas. However, as the observed responses to zero dose are not collinear with other responses, N^* and $S(y)^*$ that omit the zero tubes, replace N and $S(y)$ in Eq. (3.17).

Thus, calculate $S(y)*$ for standard and Sample A:

$$S(y)* = 122 + 276 + 133 + 306 = 837$$
$$T_S = 122 + 2(276) = 674$$
$$T_A = 133 + 2(306) = 745$$
$$S(T_i) = 674 + 745 = 1419$$

Substitute in the general forms of the Bliss equations the following values, which refer to standard and Sample A only and exclude the zero control: $f = 4$, $k = 2$, $h' = 0$, $N* = 16$. Equation (3.17) becomes

$$a' = \frac{2[(2 \times 2) + 1]S(y)* - 6S(T_i)}{16(2 - 1) + 3(0)(2 + 1)} = \frac{5S(y)* - 3S(T_i)}{8}$$

so that a' is evaluated as

$$a' = \frac{5(837) - 3(1419)}{8} = -9$$

Equation (3.18) becomes

$$b_i = \frac{3}{(2 \times 2) + 1}\left[\frac{2T_i}{(4 \times 2)(2 + 1)} - a'\right] = \frac{T_i - 12a'}{20}$$

so that the two values of b_i are evaluated as

$$b_S = (674 + 108)/20 = 39.10 \qquad \text{and} \qquad b_A = (745 + 108)/20 = 42.65$$

By Eq. (3.5),

$$R = 42.65/39.10 = 1.098$$

and sample potency is

$$\frac{1.098 \times 100 \times 50 \times 500 \times 0.4427}{5 \times 5 \times 0.4378} = 111 \quad \mu g/tablet$$

Having used this example to demonstrate the simple five point design, potencies are now calculated for all three samples appropriately as a multiple assay. Omitting responses to zero dose tubes, the following values are substituted in the general forms of the Bliss equations: $f = 4$, $k = 2$, $h' = 0$, $N* = 32$, followed by the individual values of T_i, which are

$$T_S = 122 + (2 \times 276) = 674$$
$$T_A = 133 + (2 \times 306) = 745$$
$$T_B = 118 + (2 \times 264) = 646$$
$$T_C = 125 + (2 \times 290) = 705$$

and their total $S(T_i) = 2770$ and also $S(y)* = 1634$. The general form of Eq. (3.17) becomes

$$a' = \frac{2[(2 \times 2) + 1] \times S(y)^* - 6S(T_i)}{32(2 - 1)} = \frac{5S(y)^* - 3S(T_i)}{16}$$

$$= \frac{(5 \times 1634) - (3 \times 2770)}{16} = -8.75$$

Equation (3.18) in its general form is exactly as for the simple assay, that is,

$$b_i = \frac{T_i - 12a'}{20}$$

so that

$$b_S = \frac{674 - (12)(-8.75)}{20} = 38.95$$

In exactly the same way, the other values of b_i are calculated to be

$$b_A = 42.50, \qquad b_B = 37.55, \qquad b_C = 40.50$$

Potency ratios of test solutions are obtained by Eq. (3.5) as

$$A : S = 42.50/38.95 = 1.091$$
$$B : S = 37.55/38.95 = 0.964$$
$$C : S = 40.50/38.95 = 1.040$$

Corresponding sample potencies were calculated from weighings and dilutions as has already been shown in the case of the graphically obtained potency ratios. They were

sample A   110 μg/tablet
sample B   103 μg/tablet
sample C   437 μg/tablet

It is reasonable to suppose that these are better estimates than those obtained graphically as the arithmetical method removes the element of human judgment. As the multiple assay derives a' from a larger number of zones than does the simple five point assay, it probably leads to a marginally better result.

References

Bessel, C. J., and Shaw, W. H. C. (1960). *Analyst* **85**, 389.

Bird, O. D., and McGlohon, V. M. (1972). *In* "Analytical Microbiology" (F. W. Kavanagh, ed.), Vol. II, pp. 409–436. Academic Press, New York.

Bliss, C. I. (1952). "The Statistics of Bioassay." Academic Press, New York.

Cooperman, J. M. (1972). *In* "Analytical Microbiology" F. W. Kavanagh, ed.), Vol. II, p. 454. Academic Press, New York.

Kavanagh, F. W., ed. (1963). "Analytical Microbiology," Vol. I. Academic Press, New York.
Kavanagh, F. W., ed. (1972). "Analytical Microbiology," Vol. II. Academic Press, New York.
Kavanagh, F. W. (1973). Personal communication.
Önal, Ü. (1971). Personal communication.
Skeggs, H. K. (1963). *In* "Analytical Microbiology" (F. W. Kavanagh, ed.), Vol. I, pp. 552–565. Academic Press, New York.
Tsuji, K., Elfring, G. L., Crain, H. H., and Cole, R. J. (1967). *Appl. Microbiol.* **15**, 363.
Tsuji, K. (1973). Personal communication.
Wood, E. C. (1946). *Analyst* **71**, 1.

CHAPTER 4

TUBE ASSAYS FOR ANTIBIOTICS

4.1 General Principles

Practical details of antibiotic tube methods have already been outlined in Chapter 1 using the example of the assay of neomycin. As in the case of vitamin assays, the true response is taken to be the number of cells that are present in each tube at the end of the incubation period. The nature of the response is of course opposite, increasing doses of antibiotic causing a reduction in the final numbers of cells in each tube as compared with a zero dose control.

Apart from the interaction between antibiotic and test organism, there are other factors influencing the final size of the bacterial cell population. Some of these could influence sample and standard to a different extent, thus leading to a bias in the finally estimated potency of sample. Other influences affecting sample and standard tubes equally, while not resulting in a biased potency estimate, may cause inconvenience by modifying the shape of the response curve and median response (see Section 4.1(e)) so that alternative calculation procedures have to be devised or perhaps responses to extreme doses have to be excluded from the computation of potency.

Factors that may bias the estimated potency include items (a) and (b) of the following:

(a) *The possible presence of other inhibiting substances*: Perhaps other antibiotics that occur naturally in admixture or compounded in a pharmaceutical dosage form or animal feed supplements in the case of the sample but that are not present in the standard.

(b) *The possible presence of nutrient substances in the sample, which by enriching the medium could increase growth rate leading to a low bias*: These might arise, for example, from sugars or vitamins in pharmaceutical formulations and animal feed supplements or perhaps from unconsumed nutrients in samples from an antibiotic fermentation. In the latter case it is more likely to be a problem in the earlier stages of a fermentation when potency is low and unused nutrients high. Such biases may be detected by assay at more than one dose level.

(c) *Inadequately cleaned tubes may be contaminated with absorbed traces of detergent or chromic acid*: This may cause obviously erratic responses, which would be rejected. If, however, the effect were not so large as to be obvious and if the affected tubes were not evenly distributed between standard and sample, then this could lead to bias. Clearly this is an influence to be avoided by meticulous care in the cleaning of tubes.

(d) *Time and temperature of incubation are critical*: This can be illustrated by calculations using the growth rate data for *Escherichia coli* reported by Garrett and Wright (1967). A generation time of 20 minutes at 37°C corresponds to an increase in concentration of 3% each minute. Two tubes inoculated with the same number of cells N_0 but incubated at 37 and 38°C, respectively, would differ in the number of cells as follows:

	37°C tube	38°C tube	N_{38}/N_{37}
3 hours	$178N_0$	$221N_0$	1.24
4 hours	$1005N_0$	$1340N_0$	1.33

These calculations assume a logarithmic phase of growth throughout the period of incubation. They demonstrate very clearly the importance of temperature. While differences as great as 1°C would not be expected in any single assay, a difference of say 0.2°C could be encountered if an unsuitable incubation bath were used. This would cause a difference in cell population of 5% in 3.5 hours.

It is clear that practical procedures should be designed to aim for the ideal that all tubes be rapidly brought to the same incubation temperature at the same time and that growth be terminated in all tubes at the same time.

Furthermore, to allow for any residual discrepancies in time and temperature of incubation, sample and standard tubes should be handled in such a pattern that the discrepancies tend to be self-cancelling.

The following practical techniques will help toward attainment of the ideals of uniformity of incubation conditions within an assay:

(1)   use of chilled inoculated broth (stirred well immediately before transferring to tubes) so that growth does not begin until racks of tubes are immersed in the incubation bath;

(2)   use of tubes of uniform shape and thickness of glass so that heat transfer rates may be uniform;

(3)   use of a large capacity well-stirred incubation bath so that tubes are heated at the same rate regardless of position in the bath;

(4)   termination of growth by immersion of racks in a high capacity water bath at 80°C.

To balance out residual differences in incubation conditions, each rack should include tubes representing samples and standards in a randomized arrangement.

Factors that influence sample and standard to the same extent are:

(e)   *The size and phase of growth of the inoculum*:   With a heavy inoculum the incubation period may be as low as 3 hours. A heavy inoculum
also has the effect of increasing the median response, which is defined as
the concentration of antibiotic that permits the bacterial population to
increase to 50% of the population in the zero control tube, all conditions
other than the presence of antibiotic being identical.

An inoculum in logarithmic phase of growth shortens the incubation
period. However, use of a chilled inoculum (which is in the lag phase) has
the advantage that incubation period is more easily controlled. Furthermore,
it may be possible to use the same standard inoculum over a period of several
days or more with subsequent reduction of the number of variables that
influence the form of the response line and median response.

(f)   *The nature of the medium*:   Day to day differences in batches of
apparently identical media may contribute to substantial variation. These
may be differences in pH or there may be partial decomposition of essential
nutrients of varying degree according to conditions of autoclaving and
cooling.

4.2   Response Curve—Commonly Used Forms of Expression

The general form of the response curve is illustrated qualitatively in
Fig. 4.1, in which response (cell concentration) is plotted against logarithm
of dose over a wide range. All doses below b are without effect, whereas
all doses above e exert the maximum effect, complete inhibition of growth.
The most useful working range is in the region corresponding to doses
between c and d, where the response line is steepest. An excessively steep
response line, however, would impose a restriction on the useful working
range of the assay.

The most commonly used experimental designs for turbidimetric antibiotic assays are based on interpolation from a standard curve.

Apart from the log dose versus cell concentration response line, a variety
of empirical procedures have been proposed. Some of these have been
adopted by pharmacopeias or other authoritative publications.

Proposed methods include:

(1)   Average the transmittances for each test solution, plot the mean
responses on semilog paper with doses on the log scale and responses on
the arithmetic scale, and draw a smooth curve through the points. If there
are five dose levels in geometrical progression and if the points appear to
define a straight line, then calculate ideal high and low dose responses.

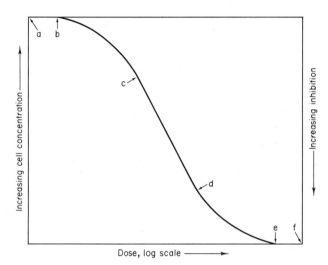

Fig. 4.1. The general qualitative form of response curve for antibiotic turbidimetric assays. Limiting responses are in the regions a to b, uninhibited growth, and e to f, complete inhibition. The central region, c to d, gives the most useful relationship between dose and response for assay purposes.

The calculation is as in the case of the plate assay for kanamycin (Example 8). This procedure was described in USP XVI but not in later editions. It is mentioned here because it is still in use in some laboratories.

(2) USP XVII suggests plotting the response as 100 minus percent transmittance against log dose. This is not really different from the method suggested in (1). The curve has the same characteristics but the direction of slope is reversed. However, the pharmacopeia goes on to suggest that if by this method a linear response is not obtained, then absorbance should be converted to percent reduction in growth and thence to probits (see also Section 4.3).

(3) Plot absorbance against log dose. This is recommended for most assays by the United States Code of Federal Regulations. Doses in geometrical progression are specified. Although not defined in the pharmacopeia, dose ratios used should be small enough to confine responses to a linear range. Dose ratios ranging from 1.14 : 1 to 1.4 : 1 are specified in assays described in the CFR.

(4) Plot relative cell concentration, also known as corrected absorbance, against log dose as illustrated in Fig. 4.1 (see Section 3.2).

(5) USP XVIII suggests either three dose levels of standard and sample in geometrical progression or a five dose level standard curve with one dose level of sample. It does not specify what function of response is to be used but gives calculation procedures that are applicable when some function

of response plots linearly against logarithm of dose. In addition to factorial tables such as given in Appendix 2., USP XVIII gives tables that are applicable when there are three or four doses in the ratios included in the sequence 1.5, 2.0, 3.0, and 4.0. However, as these calculations are based on a linear response to log dose, it seems more logical to use a dose series forming a geometrical progression. This is more efficient as the responses are evenly spaced. The factorial tables are also simpler.

(6)   The European Pharmacopeia, Volume II, referring to both diffusion and turbidimetric assays requires that dose–response lines for standard and sample be shown to be both linear and parallel within a given probability level. It goes on to say that the method of calculation would depend on the optical apparatus used, but a suitable transformation of the response should be found so as to give linearity of the dose–response lines. The pharmacopeia does not make any specific suggestions of suitable transformations for this type of assay nor does it give an example of a calculation.

(7)   Plot transmittance against dose.

These varied proposals arose at times when theory was nonexistent. The adaptation of calculation routine that had worked satisfactorily in agar diffusion assays lacks any logical basis. Use of transmittance directly and plotting against dose or log dose results in strongly curved lines. When random errors of the points is substantial, fitting a good curve becomes a matter of luck. It is more logical to plot optical density, which may approximate to relative cell concentration. A plot of optical density versus log dose may be linear, or almost so, over part of the range of doses.

4.3   Linearization of Dose–Response Relationships—
Theoretical Considerations

The standard curve procedures outlined in Section 4.2 are widely used and with apparent success. They have certain inherent disadvantages, however:

(1)   Subjective assessment of the straight line of best fit or the free hand drawing of a curve introduces errors or biases of unknown size; evaluation from the internal evidence of such assays is not possible.

(2)   The absence of any logical basis for defining a mathematical relationship between dose and response hampers efforts to identify ideals in assay design.

Attempts to derive a consistently reliable straight line relationship between function of dose and function of response appear to have met with only partial success.

Growth-inhibiting substances may influence the final concentration of microorganisms in an assay tube by three possible separate mechanisms or by a combination of two or all of these. The possible mechanisms are (1) increasing the generation time, (2) increasing the lag time, (3) killing a fraction of the organisms.

Equation have been derived to represent mechanisms (1) and (2) either singly or combined by Kavanagh (1968). Most of those cases of interaction between an organism and an inhibitor that have been studied depend on mechanism (1). In such cases, and when the inoculum is in the log phase of growth, according to Kavanagh (1975a) all influences on growth of the organism may be represented by the general equation

$$N_t = N_0 \exp(k_0 + f(v)k_m - k_a C)t \tag{4.1}$$

where N_0 is the initial concentration of living cells, N_t the concentration of cells at time t, C the concentration of inhibitor, k_0 the generation constant in the absence of inhibitor, k_a the inhibitory coefficient of the inhibitor, k_m the cofficient for menstruum effect, and $f(v)$ the function of the volume of test solution added to the assay tube.

The menstruum coefficient represents the effect of substances other than the inhibitor to be determined, which may also be present in the test solution. These may be either inhibitors (k_m is negative) or growth-promoting substances (k_m is positive), or ions reducing the activity of the antibiotic to be determined (k_m is positive). The menstruum coefficient k_m may be a composite figure comprising all three effects.

Thus the term $f(v)k_m$ represents the effect of all interfering substances in the test solution.

Although the general Eq. (4.1) cannot be written exactly to describe an individual assay, a study of its form is useful for an understanding of principles. It may be rewritten first as

$$\ln N_t = \ln N_0 + k_0 t + f(v)k_m t - k_a C t \tag{4.2}$$

then substituting (i) for the terms unaffected by concentration of antibiotic to be assayed,

$$A = \ln N_0 + k_0 t + f(v)k_m t \tag{4.3}$$

and (ii) for the term that is dependent on concentration of antibiotic to be assayed,

$$B = k_a t \tag{4.4}$$

it becomes

$$\ln N_t = A - BC \tag{4.5}$$

It follows that if (i) the initial concentration of inoculum is the same in all tubes, (ii) all tubes undergo identical treatment as regards time and temperature of incubation and termination of growth, and (iii) $f(v)k_m t$ is the same for all tubes, then a plot of log N_t versus C should be a straight line. It also follows that the straight lines for a series of samples should all converge at a common origin corresponding to the response to the zero control tube. This provides the basis for a slope ratio assay design.

Conditions (i) and (ii) are simply attained by good technique and strict attention to detail. Condition (iii) however may be more difficult to achieve. If interfering substances are present in significant amounts in the "unknown" test preparation but not in the standard, then the quantity of interfering substance will vary in direct proportion to the dose and will lead to a biased potency estimate.

Possible remedies include preliminary purification of the sample or, if the quantity and nature of the interfering substance is known (e.g., a solvent used to extract an antibiotic), its effect may be compensated by ensuring that every tube contains the same quantity of the interfering substance.

Kavanagh suggests that Eq. (4.5) may be an appropriate representation of the responses to penicillin, erythromycin, tylosin, chloramphenicol, and tetracycline. In Fig. 4.6b, it is seen that in the assay of tetracycline using *Staphylococcus aureus* as test organism, this equation gives a good straight line response up to a concentration of about 0.30 IU/tube (see Example 15).

Assays of streptomycin using *Klebsiella pneumoniae* do not follow this relationship, but a plot of log N_t versus C^2 gives a straight line, as shown in Fig. 4.5g. This relationship is impracticable for routine assays.

For inhibition mechanisms more complex than increase in generation time alone, expressions similar to Eq. (4.5) but containing additional terms are derived. It appears that in these cases too, Eq. (4.5) may sometimes be a close enough approximation.

The third of the postulated mechanisms of inhibition is that the active substance kills a fraction of the cells in the inoculum but permits the remainder to grow at the same rate and with the same lag time as those in the zero control tube. If this were so, and if it were due to varying sensitivities of individual cells, it would seem reasonable to suppose that the sensitivities of individuals would be normally distributed with respect to some function of dose. In this event, then a plot of N/N_{max} against an appropriate function of C should be in the form of an inverted cumulative normal distribution, where N_{max} is the value of N in the zero control tube. Expressed in another way, a plot of N/N_{max} on a probability scale against the appropriate function of C should be a straight line. Although in fact there appear to be few examples of this mode of action, it is found that in many cases such a plot (probability) of N/N_{max} versus log C does closely approximate to a straight line. In the case of assays of streptomycin using *K. pneumoniae*, however, a probability plot of N/N_{max} against concentration itself is found to give a straight line. Such linearization procedures were first proposed independently by Hemmingsen (1933) and Gaddum (1933) for use in circumstances when response is quantal. That is, response is measured as the number (or percentage) of subjects in a group receiving a certain treatment (dose of a preparation) that show a positive effect. The effect is of the "all or none" type and may be, for example, the death of an individual animal.

In toxicity tests on mice, Hemmingsen and Gaddum used the normal distribution curve in its cumulative form. Fifty percent deaths in a group corresponded naturally to zero deviation from the mean of all possible responses. Other percentage responses were expressed in terms of "normal equivalent deviation" from this mean. A straight line was obtained on plotting responses (deaths in a group) converted in this way to normal equivalent deviation (standard deviation) against logarithm of the dose.

To avoid the use of negative numbers, Bliss (1934a,b) added 5.00 to all values of the normal equivalent deviation and gave the new values the name "probability unit," which was shortened to "probit."

The relationship between these units and the normal distribution curve is shown in Table 4.1 and also graphically in Fig. 4.2.

Table 4.1

Relationship between Percentage Response,
Normal Equivalent Deviation, and Probit

Percentage response	Normal equivalent deviation	Probit
2.28	−2.0	3.0
15.87	−1.0	4.0
50.00	0.0	5.0
84.13	+1.0	6.0
97.72	+2.0	7.0

An alternative procedure intended for the processing of quantal data is the angular transformation, defined by

$$p = \sin^2 \phi \qquad (4.6)$$

where p is the proportionate response (percentage of subjects reacting) and ϕ an angle between 0° and 90°.

Berkson (1944) describes the logit transformation that may be used for quantitative responses, i.e., when *individual* responses may vary between 0 and 100%. This is the case in turbidimetric assays where final cell con-

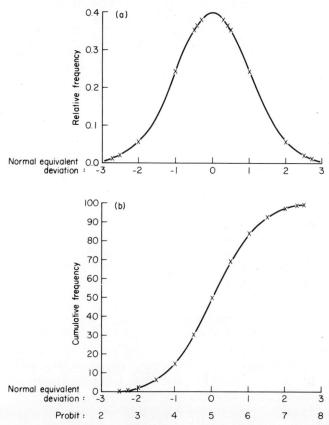

Fig. 4.2. Illustrations of the normal distribution. Graph (a) shows the normal distribution curve. The vertical axis indicates the relative frequency of observed values in a normally distributed population according to their deviation from the mean. The total area under the curve is unity. The areas between ± 1 and ± 2 normal equivalent deviations are 0.683 and 0.955, respectively, of the whole area. Graph (b) indicates the area under curve (a) lying to the left of any vertical line for the corresponding normal equivalent deviation. It is thus a cumulative frequency curve.

centration in an individual tube may be any value between zero and that maximum as displayed by the zero dose control tube. The logit (y) is defined by Berkson as

$$y = \ln(p/q) \tag{4.7}$$

where p is proportionate response and $q = (1 - p)$.

Fisher and Yates (1963), however, use essentially the same transformation differing by a factor of 2 and define the logit (z) as

$$z = 0.5 \ln(p/q) \tag{4.8}$$

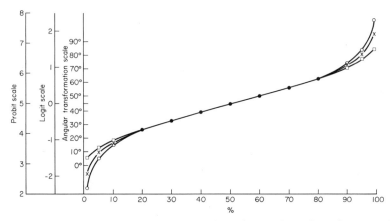

Fig. 4.3. A comparison of the probit, logit, and angular transformations of response. The *x* axis indicates response as a percentage of maximum response. Corresponding values of the probit, logit, and angular transformations of response are plotted on the *y* axis on individual scales chosen so as to demonstrate the almost indistinguishable effect of these three functions at values of *x* between 20 and 80%. Probit, × ; logit, ○; angular, □.

In fact, all three of these transformations have very similar effects in the response range 20–80%. This can be seen in Fig. 4.3, where all three have been plotted on scales chosen to show the virtual identity of the transformations in that range. In a well-planned assay, most responses fall within this range and so it seems to be only a matter of convenience which transformation should be selected.

If a statistical evaluation is to be carried out, then the angular transformation is stated to have the advantage that all observations carry equal weight and so the complication of weighting coefficients is avoided. Bearing in mind the virtual coincidence of the three transformations in the range 20–80%, it is clear that weighting could be omitted for any of the relationships in this range.

In turbidimetric assays using the angular transformation of response, p is the concentration of cells relative to the concentration in the zero dose control tube, i.e., N/N_{max}. The conversion is made very conveniently from tables such as that of Fisher and Yates (1963). This transformation approximates closely to the probit transformation in the range $p = 0.07$ to 0.93. According to Bliss, it is justified in general by statistical convenience rather than any biological model.

If responses are measured in a nephelometer using the zero dose control to set the instrument 100% reading, then sample responses are obtained directly as p. This is also equally convenient if the probit transformation is to be used.

Another completely empirical method suggested by Kavanagh (1972) is the "inverse log plot." In this procedure, responses are plotted on semilog paper so that, for example, a response of 2 is plotted on line 8, a response of 3 on line 7, etc. The procedure is then repeated but shifting the points one unit so that a response of 2 is plotted on line 7, a response of 3 is plotted on line 6, etc. The procedure is repeated at varying starting points and a family of curves is obtained. The lines show graded differences in curvature and the one approximating to a straight line is selected for use.

This method was designed with the intention of providing a straight line for interpolation of sample potencies by computer.

To summarize, the curve-straightening procedures that have been described include the following relationships:

(1)   the plot of log N versus dose, which is based on the equation log $N = A - BC$,

(2)   the probit* transformation of response versus dose or log dose,

(3)   the logit transformation of response versus dose or log dose,

(4)   the angular* transformation of response versus dose or log dose,

(5)   the inverse log plot.

4.4   Dose–Response Linearization Procedures in Practice

The relative merits of some of these graphical representations may be seen by plotting the same data in different ways. For the following illustrations data have been taken from routine assays carried out in several laboratories. Two of these were assays using an empirical standard curve method of computation. It will be shown in Sections 6.14 and 7.4 how the same data may be processed arithmetically to give the best estimate of sample potency. However, when the assay is designed for graphical interpolation, the purely arithmetical computation is often very tedious.

An assay of streptomycin in which the test organism was stated to be *Klebsiella pneumoniae* (Example 13) is illustrated by Fig. 4.4. The intended method of sample potency estimation was by reading sample responses at three dose levels from a seven dose level standard curve. In this case standard and sample response lines are so close that only the standard line is shown in each of the four representations of Fig. 4.4.

Figure 4.4a, transmittance versus dose, and Fig. 4.4b, absorbance versus dose, are both sigmoid and are of use only in an interpolation method. In all

---

* Abridged tables adapted from those of Fisher and Yates are given in Appendices 7 (for probit) and 8 (for angular transformations of response).

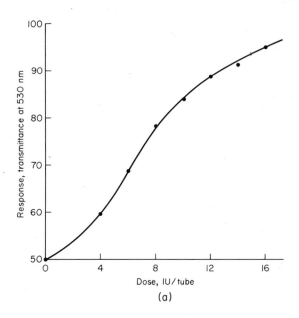

(a)

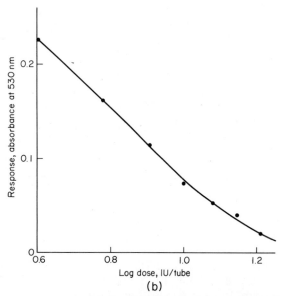

(b)

Fig. 4.4. Graphical representation of the dose–response relationship in the turbidimetric assay of streptomycin (Example 13) using *Klebsiella pneumoniae*. (a) The observations as originally recorded and used, transmittance versus dose. (b) Observations expressed as absorbance versus log dose. (c) Observations expressed as log absorbance versus dose. (d) Responses transformed to corresponding probit and plotted versus log dose.

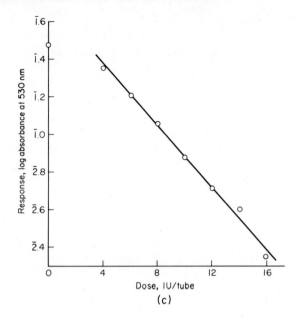

(c)

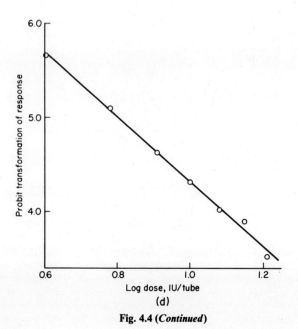

(d)

Fig. 4.4 (*Continued*)

these representations the responses to doses 14 and 16 U/ml deviate somewhat from the best smooth curve or best straight line. This seems to suggest
that a fault in technique has biased these observations. Apart from these
discrepancies both Figs. 4.4c, log absorbance versus dose, and 4.4d, probit
(based on absorbance) versus log dose, illustrate good straight line relationships. It should be noted, however, that in Fig. 4.4c the response to zero
dose does not lie on the same straight line as other observations. This is
possibly because growth has continued in the log phase in tubes containing
antibiotic but not in the zero control tubes. Potency may be calculated
using the relationship shown in Fig. 4.4c as a slope ratio assay omitting
response to zero dose, or alternatively using the relationship of Fig. 4.4d
as a parallel lines assay. The latter is shown complete with statistical evaluation in Chapter 6. Details of this assay and graphical interpolation of sample
potency are given in Section 4.5, Example 13.

Another assay of streptomycin also reported as using *K. pneumoniae*
(Example 14) is illustrated in Fig. 4.5. These two streptomycin assays were

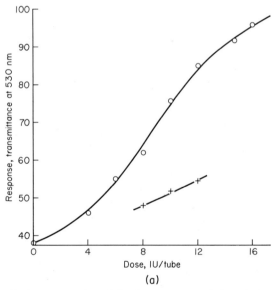

Fig. 4.5. Graphical representation of the dose–response relationship in the turbidimetric
assay of streptomycin (Example 14) using *K. pneumoniae*. (a) Observation as originally recorded
and used, transmittance versus dose. (b) Observations expressed as absorbance versus log dose.
(c) Observations expressed as log absorbance versus dose. (d) Observations transformed to
probit and plotted versus log dose. (e) Observations transformed to probit and plotted versus
dose. (f) Observations transformed to corresponding angle by means of the relationship $p =
\sin^2 \phi$ and plotted versus dose. (g) Observations expressed as log absorbance versus (dose)^2.

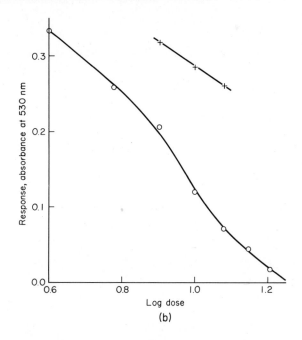

(b)

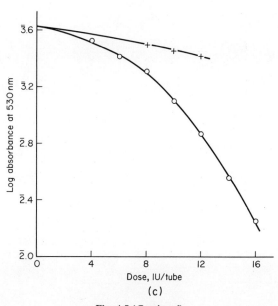

(c)

Fig. 4.5 (*Continued*)

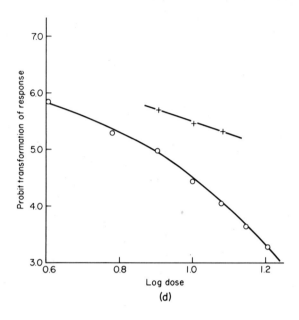

(d)

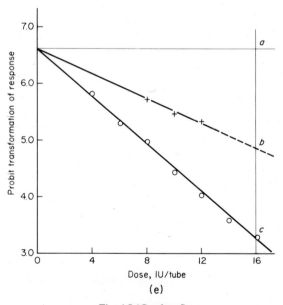

(e)

Fig. 4.5 (*Continued*)

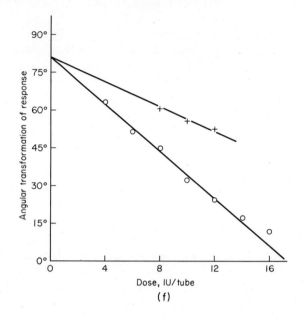

Dose, IU/tube

(f)

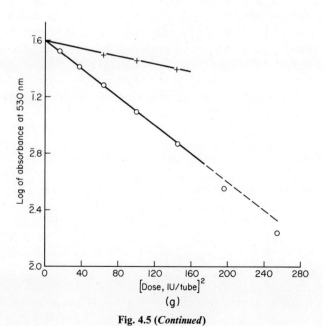

[Dose, IU/tube]^2

(g)

Fig. 4.5 (*Continued*)

carried out in different laboratories in 1968 and although conditions of test were nominally the same, the nature of the response lines was different.

The discrepancy was unexplained at that time. Now it is known that the response line form of Example 13, Fig. 4.4c, is typical of many assay systems and is in conformity with Eq. (4.5). Assays of streptomycin using *K. pneumoniae* are an exception, however, in that a straight line is given by a plot of log N_t versus C^2, compare Fig. 4.5g. It seems very likely then that the organism used in the assay (Example 13) was in fact *not K. pneumoniae*.

In this particular assay, as sample and standard were qualitatively similar, use of the wrong organism would probably not lead to bias. In the case of assay of a sample containing additional ingredients, however, bias is possible. (See also Section 1.2.)

The importance of checking that the organism is in fact what is stated on the label will now be clear.

This second assay (Example 14) too was designed for sample responses at three dose levels to be interpolated from a seven dose level standard curve. In this case, however, the response lines are far apart and so both can be illustrated. Graphs of Figs. 4.5a, transmittance versus dose, and 4.5b, absorbance versus log dose, are both sigmoid and useful only for graphical interpolation. In contrast to the previous streptomycin assay, response lines of Fig. 4.5c, log absorbance versus dose, are strongly curved. A plot of probit based on absorbance versus log dose is also curved, (Fig. 4.5d). However, a plot of the same probits versus dose gives two straight lines with a common zero at zero dose, (Fig. 4.5e). An angular transformation of response (absorbance) is shown in Fig. 4.5f, in which it will be noted that the standard response line is curved in the range of the highest doses 14 and 16 IU/tube. This deviation is not unexpected as the responses are, respectively, on the limit of and beyond the useful working range of this transformation (i.e., 7–93% of the absorbance of the zero control tube).

Sample potency is obtained graphically using the probit transformation in Example 14. Again, as the assay was not designed with a probit transformation of response in mind, the calculation of sample potency by purely arithmetical means is very tedious and is not worthwhile when only a potency estimate without confidence limits is required.

To obtain a potency estimate with confidence limits it is preferable to use the angular transformation, thus avoiding the necessity for weighting. This calculation is shown in Chapter 7, where observations corresponding to doses 14 and 16 IU/tube are discarded as not being collinear with other observations.

A balanced assay designed for purely arithmetical computation of sample potency is illustrated in various forms in Fig. 4.6. This is an assay of a pharmaceutical form of tetracycline. Both standard and sample are at six dose

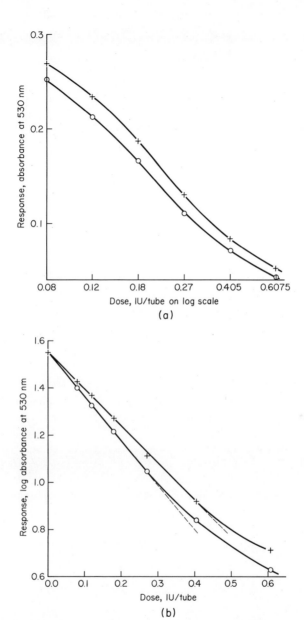

Fig. 4.6. Graphical representation of the dose–response relationship in the turbidimetric assay of tetracycline (Example 15) using *Staphylococcus aureus.* (a) Sigmoid response lines of absorbance versus log dose. (b) A plot of log of absorbance versus dose, which is linear up to dose 0.27 IU/tube. This is used in the graphical estimation of potency in Section 4.8. (c, d) These show, respectively, probit and angular transformation of response plotted versus log dose. Both approximate closely to linear relationships in the central dose range (0.12 to 0.405 IU/tube). In Chapter 6 potency and confidence limits are estimated using the relationship illustrated in (d).

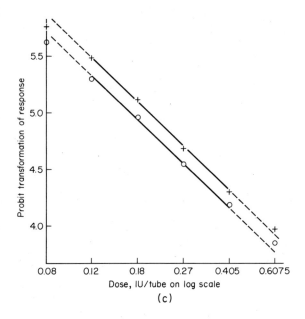

(c)

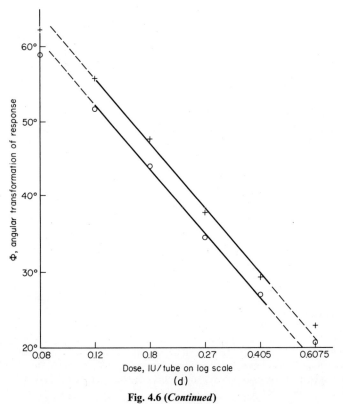

(d)

Fig. 4.6 (*Continued*)

levels in geometrical progression so that the potency ratio may be calculated by means of a·probit or angular transformation of optical density versus log dose.

The probit transformation of absorbance illustrated in Fig. 4.6c has been reasonably successful, although there is evidence of deviation from linearity of the responses to extreme doses, which indicates a slightly sigmoid curve. The angular transformation of response is shown in Fig. 4.6d. In Section 4.7 it is shown how potency ratios may be obtained from this latter relationship. Both these transformations of response versus log dose relationships lead to computation procedures similar to those used in the already familiar agar diffusion assays, which are also dependent on parallel response lines.

The curvature of an absorbance versus log dose line is shown clearly in Fig. 4.6a. In Fig. 4.6b it is seen that Kavanagh's theoretical relationship, log response versus dose, is valid in the dose range 0.08–0.27 IU/tube. Two straight lines intersect at a point corresponding to growth in the zero dose control tube (at doses 0.405 and 0.6075 IU/tube, however, the lines are markedly curved). Using only the linear part of the response line and treating as a slope ratio assay, potency is estimated graphically in Section 4.8.

4.5 Potency Estimation by Interpolation from a Standard Curve

In the assay of a streptomycin sample illustrated in Example 13, potency was initially estimated by interpolation from a standard curve. This was the standard procedure in the laboratory concerned and the assay design was appropriate for this method of estimation. However, this was one of many individual assays carried out in a collaborative assay and so it was necessary to evaluate the assay and establish confidence limits for the estimated potency. A purely arithmetical procedure was devised based on the relationship angular transformation of response versus log dose, which results in parallel response lines for standard and sample.

As has already been stated, the arithmetical procedure based on a non-"purpose built" design is tedious. It would not be justifiable in normal circumstances and so is only shown in its complete form leading to confidence limits. This calculation is given in Section 6.14.

Example 13: Sample of Streptomycin Sulfate Compared with Standard Preparation of Streptomycin Sulfate

Test organism: K. *pneumoniae* ATCC 10,031
Incubation: 3 hours at 37°C
Weighings and dilutions:
Standard: potency 780 IU/mg:

62.05 mg → 50 ml : 5 ml → 50 ml (96.8 IU/ml)

Sample:

$$63.35 \text{ mg} \rightarrow 50 \text{ ml} : 5 \text{ ml} \rightarrow 50 \text{ ml}$$

These were further diluted to give the required dose in 1 ml, i.e., 4 to 16 for standard (multiplied by the factor 0.968), and nominal doses of 8 to 12 IU/ml for sample.

Observations: Transmittance measured using a Lumetron photometer at 530 nm. Observations are recorded in Table 4.2 and shown graphically in Figs. 4.4a–4.4d.
Reading the sample test solution potencies from Fig. 4.4a, potency ratios, sample test solution/standard test solution were calculated (Table 4.3). The mean potency ratio = 1.002. This should be compared with the potency ratio calculated in Chapter 7, i.e., 1.021. The latter, which does not involve human judgment of lines of best fit, must be considered the best estimate. However, for completeness, the sample potency as estimated graphically is calculated:

$$\frac{96.8 \times 1.002 \times 50 \times 50}{5 \times 63.35} = 766 \quad \text{IU/mg}$$

Table 4.2

Observations of Example 13

Dose IU/tube	log dose	Transmittance			Mean transmittance	Mean absorbance
Standard						
4	0.6021	58.5	58.0	62.0	59.5	0.226
6	0.7782	68.0	68.0	70.5	68.8	0.162
8	0.9031	78.5	79.5	76.5	78.2	0.107
10	1.0000	83.5	86.0	83.0	84.2	0.075
12	1.0792	90.0	88.5	88.0	88.8	0.052
14	1.1461	90.0	93.0	91.0	91.3	0.040
16	1.2041	93.5	96.5	96.0	95.3	0.021
Sample						
8	0.9031	78.5	80.5	77.0	78.7	0.104
10	1.0000	84.0	85.0	84.0	84.3	0.074
12	1.0792	87.5	89.0	89.0	88.5	0.053
Zero						
control	—	50.0	51.0	49.0	50.0	0.301

Table 4.3

Sample nominal potency, IU/tube	Mean transmittance	Potency from graph, IU/tube	Potency ratio
12	88.5	11.9	0.992
10	84.3	10.0	1.000
8	78.7	8.1	1.013

4.6 Graphic Estimation of Potency by Probit of Response Versus Dose

The assay of streptomycin illustrated in Example 14 was again by the routine design used in a certain laboratory for graphical interpolation of estimated potency.

To enable a statistical evaluation of the method's potential precision to be made, arithmetical procedures were devised. First the original graphical interpolation is given. It is then shown that a better graphical estimate may be made by plotting probit of response against dose.

However, for the statistical evaluation in Chapter 7 using angular transformation of response, the responses to doses 14 and 16 IU are rejected, as the line is curved in this region.

Example 14: Pharmaceutical Preparation Containing Streptomycin Compared with Standard Streptomycin Sulfate

Test organism: *Klebsiella pneumoniae* ATCC 10,031
Incubation: 3 hours at 37°C
Dilutions:
Standard streptomycin sulfate, 776 IU/mg:

$$54.3 \text{ mg} \quad (42,200 \text{ IU}) \rightarrow 42.2 \text{ ml} \quad (1000 \text{ IU/ml})$$

Sample, nominal potency 485 IU/mg:

$$106.2 \text{ mg} \rightarrow 100 \text{ ml} \quad (\text{nominal } 1000 \text{ IU/ml})$$

Both primary dilutions were diluted further to give the required doses in the 1 ml of test solution added to each tube.

Observations: Transmittance (as %) by Lumetron photometer at 530 nm. Observations and their transformation to probits are shown in Table 4.4. Various representations of dose–response relationships are given in Figs. 4.5a–4.5g.

Reading the sample test solution potencies from the graph (Fig. 4.5a) potency ratios are calculated as shown in Table 4.5. The mean relative potency is 0.511. The sample potency is calculated as

$$\frac{1000 \times 0.511 \times 100}{106.2} = 481 \quad \text{IU/mg}$$

It was found that with a little more effort, use of a probit transformation of response made it posssible to obtain a straight line response and thus obtain an estimate of potency that although still graphical did not involve the hazards of drawing curves of best fit.

In Fig. 4.5e probits are plotted against dose of streptomycin and two lines intersecting at zero dose drawn as judged by eye to be the best fit. Observations modified in this way fulfil the requirements of a slope ratio assay. Distances between the intercepts *a*, *b*, and *c* on a vertical line (drawn at dose

Table 4.4

Data of Example 14[a]

Dose, IU/tube	Transmittance, T				T, mean	Absorbance A	$100A/A_{max}$	Probit
Standard								
4	45.5	44.0	47.5	48.0	46.3	0.334	79.6	5.83
6	54.0	55.5	55.0	56.0	55.0	0.260	62.0	5.31
8	62.0	62.5	61.5	62.5	62.1	0.207	49.3	4.98
10	76.0	76.5	74.5	75.0	75.5	0.122	29.1	4.45
12	84.0	85.0	85.5	85.0	85.0	0.071	16.9	4.04
14	92.0	92.5	91.5	92.0	92.0	0.036	8.6	3.63
16	96.0	97.5	95.5	94.5	95.9	0.018	4.3	3.28
Sample								
8	47.0	47.0	49.0	49.0	48.0	0.319	76.0	5.71
10	52.0	50.5	51.0	53.5	51.8	0.286	68.1	5.47
12	52.5	57.5	54.0	54.0	54.5	0.264	62.9	5.33
Zero control	37.0	38.5	38.0	38.5	38.0	0.420	100.0	(8.72)

[a] Observations, transmittance at 530 nm using a Lumetron photometer, together with transformation of these responses to the corresponding probit.

Table 4.5

Sample nominal potency, IU/tube	Mean transmittance	Potency from graph	Potency ratio
8	48.0	4.27	0.534
10	51.8	5.11	0.511
12	54.5	5.87	0.489

16 IU/tube) by the two response lines and a horizontal line at the level of response to zero dose were measured and used to calculate the slope ratio.

Ideally the response lines should intersect one another at the y axis (zero dose) at a probit corresponding to zero response, that is, 100% growth. In this case they intersect at a probit value of 6.60, which corresponds to 94.5% growth. Bearing in mind that probits decrease in reliability at increasing distances upward or downward from 5.0, this deviation from the ideal may be considered quite acceptable.

The potency ratio from Example 14 can now be calculated from a plot of probit against dose.

The slope ratio is given by

$$(b - a)/(c - a) = (4.84 - 6.60)/(3.26 - 6.60) = -1.76/-3.34 = 0.527$$

Thus the sample primary dilution of nominal concentration 1000 IU/ml is shown to have an actual concentration of only 527 IU/ml. This leads to a sample potency of

$$\frac{527 \times 100}{106.2} = 496 \quad \text{IU/mg.}$$

As in the case of Example 12 the purely arithmetical procedure cannot be justified for routine work.

4.7 Arithmetical Estimation of Potency from an Assay of Balanced Design by Angular Transformation of Response

The third example of an antibiotic turbidimetric assay was designed as a balanced assay. In Example 15 there are six dose levels in the ratio 3:2 covering the range 0.0800 to 0.6075 units of tetracycline per tube. It was intended that the observations should provide the basis for an assay and also show the most useful working range for future work. If necessitated by failure to conform to a straight line relationship, responses to extreme doses could be discarded yet still leave the basis for a balanced assay.

The probit and angular transformations of dose gave the relationships illustrated in Figs. 4.6c and 4.6d, respectively. The latter was taken as the basis of calculation using (i) only dose levels 0.1200 to 0.4050 IU/tube, and (ii) all six dose levels. It is thus demonstrated that the slight deviations from linearity by the extreme responses have a quite negligible effect on the calculated potency when using this relationship, which is based on a function of response versus log dose.

It was also found that the relationship log response versus dose was applicable over the dose range zero to 0.2700 IU/tube and so a graphical estimate of potency based on slope ratio is also illustrated.

For the estimate of confidence limits, the angular transformation of response is used again in Chapter 6.

Example 15a: Pharmaceutical Preparation of Tetracycline (Oral Suspension) Assayed against Standard Tetracycline Hydrochloride

Test organism: *Staphylococcus aureus* ATCC 6538-P
Incubation: 3.5 hours at 37°C
Weighings and dilutions:
Standard, potency 967 IU/mg:

51.9 mg → 50.2 ml : 5 ml → 50 ml : 5 ml → 50 ml : 10 ml → 100 ml

giving a solution of potency 1 IU/ml.
Sample, contains nominally 125 mg tetracycline base in each 5 ml dose:

5 ml → 250 ml : 10 ml → 50 ml : 5 ml → 50 ml : 10 ml → 100 ml

giving a solution of nominal potency 1 IU/ml.

Both standard and sample solutions were then further diluted as follows:

2 ml → 25 ml	giving	0.08 IU/ml
3 ml → 25 ml	giving	0.12 IU/ml
4.5 ml → 25 ml	giving	0.18 IU/ml
6.75 ml → 25 ml	giving	0.27 IU/ml
8.10 ml → 20 ml	giving	0.405 IU/ml
12.15 ml → 20 ml	giving	0.6075 IU/ml

Observations and their transformation to the corresponding angle are shown in Table 4.6.

Calculation of potency using values of ϕ for four dose levels only, i.e., 0.120 to 0.405 IU/tube:

$$E = \tfrac{1}{20}[3(27.1 + 29.5) + (34.8 + 38.1) - (44.1 + 47.7) - 3(51.9 + 55.8)]$$
$$= -172.2/20 = -8.61$$

$$F = \tfrac{1}{4}[(55.8 + 47.7 + 38.1 + 29.5) - (51.9 + 44.1 + 34.8 + 27.1)]$$
$$= 13.2/4 = 3.30$$

$$I = \log 1.5 = 0.1761$$

$$b = E/I = -8.61/0.1761 = -48.89$$

$$M = F/b = 3.30/-48.89 = -0.0675 = \bar{1}.9325$$

Potency ratio:

$$\text{antilog } \bar{1}.9325 = 0.8561$$

Sample potency:

$$\frac{1.0 \times 0.8561 \times 100 \times 50 \times 50 \times 250 \times 1}{10 \times 5 \times 10 \times 5 \times 1000} = 21.4 \quad \text{mg/ml of tetracycline base}$$

This corresponds to 107 mg of tetracycline base per 5 ml dose.

Using values of ϕ for all six dose levels, potency is calculated as follows:

$$E = \tfrac{1}{70}[5(20.8 + 23.0) + 3(27.1 + 29.5) + (34.8 + 38.1)$$
$$- (44.1 + 47.7) - 3(51.9 + 55.8) - 5(62.3 + 59.1)]$$
$$= -560.2/70 = -8.003$$

$$F = \tfrac{1}{6}[(62.3 + 55.8 + 47.7 + 38.1 + 29.5 + 23.0)$$
$$- (59.1 + 51.9 + 44.1 + 34.8 + 27.1 + 20.8)]$$
$$= 18.6/6 = 3.10$$

$$b = E/I = -8.003/0.1761 = -45.44$$

$$M = F/b = 3.10/-45.44 = -0.0682 = \bar{1}.9318$$

Table 4.6

Data of Example 15[a]

Dose, IU/tube	log of 100 × dose	Transmittance			Mean transmittance	Mean absorbance A	A/A_{max}	ϕ (degrees)
Standard 0.0800	0.903	56.5	56.0	55.5	56.0	0.252	0.737	59.1
0.1200	1.079	62.0	60.5	62.0	61.5	0.212	0.620	51.9
0.1800	1.255	68.5	68.0	68.5	68.3	0.166	0.485	44.1
0.2700	1.431	77.5	77.5	77.5	77.5	0.111	0.325	34.8
0.4050	1.608	85.0	85.0	85.0	85.0	0.071	0.208	27.1
0.6075	1.784	90.5	90.5	91.0	90.7	0.043	0.126	20.8
Sample 0.0800	0.903	54.0	54.0	54.0	54.0	0.268	0.784	62.3
0.1200	1.079	58.0	58.5	58.5	58.3	0.234	0.684	55.8
0.1800	1.255	65.0	65.0	65.0	65.0	0.187	0.547	47.7
0.2700	1.431	74.5	74.0	74.0	74.2	0.130	0.380	38.1
0.4050	1.608	82.5	82.5	83.0	82.7	0.083	0.243	29.5
0.6075	1.784	89.0	89.0	88.5	88.8	0.052	0.152	23.0
Zero control	—	45.5	45.0	46.0	45.5	0.342	1.000	

[a] Observations, transmittance at 530 nm using a Spectronic 20 spectrophotometer, together with transformation of these responses to the corresponding angle ϕ by means of Eq. (4.8).

Potency ratio:

$$\text{antilog } \bar{1}.9318 = 0.8547$$

This leads to a sample potency of 21.4 mg/ml of tetracycline base as before.

4.8   Graphic Estimation of Potency Using the Relationship Log Response Versus Dose

Kavanagh's relationship log response versus dose is now applied to the data of Example 15a and the potency estimated from the graph by means of the slope ratio. Only the portions of the response lines corresponding to doses of up to 0.27 IU/tube for standard and up to 0.405 IU/tube for sample are used, this being the range of linear relationship.

It is shown in Section 8.8 that to use responses outside the range of linearity in the slope ratio calculation is liable to result in substantial bias as potency ratios depart from 1.0. This is in marked contrast to balanced parallel line assays.

The values to be plotted in the graph are shown in Table 4.7 as a continuation of Example 15a. The graph is shown as Fig. 4.6b.

Table 4.7

Tabulation of Values for the Graph Shown in Fig. 4.6b from the Data of Example 15a

Dose, IU	Response, absorbance	log (response × 100)
Standard		
0.0800	0.252	1.4014
0.1200	0.212	1.3263
0.1800	0.166	1.2201
0.2700	0.111	1.0453
0.4050	0.071	0.8451
0.6075	0.043	0.6335
Sample		
0.0800	0.268	1.4281
0.1200	0.234	1.3692
0.1800	0.187	1.2718
0.2700	0.130	1.1139
0.4050	0.083	0.9191
0.6075	0.052	0.7160
Zero control	0.342	1.5340

Example 15b: An alternative Calculation Procedure

Slope ratio may be determined from the graph with greater convenience in reading if the straight portion of each response line is extrapolated. Intercepts are then measured on a vertical line drawn at any convenient dose level. In this case a dose of 0.30 IU/tube was selected and intercepts were found as shown in the accompanying tabulation. It will be seen that the value of the intercept *a* is not identical with the value tabulated for log response to zero dose. The reason is simply that response to zero dose is, like any other response, subject to random error. Thus, for slope ratio,

$$(c - a)/(b - a) = (1.090 - 1.555)/(0.990 - 1.555) = 0.823$$

		Intercept	log response
(a)	Horizontal line through common origin		1.555
(b)	Standard response line		0.990
(c)	Sample response line		1.090

Thus potency ratio of test solutions is 0.823, which leads to a potency estimate for the sample of 20.6 mg tetracycline base/ml.

These same observations were graphed independently by an experienced analyst who concluded that the potency ratio of test to standard solution was 0.868. The limitations of the graphical method are thus illustrated.

Finally, by fitting multiple regression equations the potency ratio was calculated as 0.840. This assay was not designed for computation of potency estimate by the slope ratio method. The uneven spacing of doses on a linear scale makes such a computation unwieldy. Although the result of fitting multiple regression equations has been quoted, this method cannot be recommended in these circumstances and so is not reproduced here. Nevertheless the value 0.840 must be considered the best estimate of potency that can be obtained from the data corresponding to the straight line portion of these log response–dose graphs. If the log response–dose relationship were to be used in future work, then convenient doses would be either the series 0.06, 0.12, 0.18, and 0.24 or 0.09, 0.18, and 0.27 IU/ml thus permitting potency estimation by the simpler arithmetical procedures described in Sections 3.7–3.9.

It should be noted that Kavanagh does not advocate the slope ratio method for use with the log response versus dose relationship, lest it should be applied indiscriminately as a calculation procedure in cases where there are deviations rendering it invalid. Instead he uses a computer to interpolate from a standard curve. Any drift in potency estimated at different dose levels is taken as a warning of possible invalidity necessitating investigation. However, the problem of misuse of convenient calculation procedures is general and not just a feature of this particular relationship. It is always necessary to inspect the data to see if they conform to the basic assumptions implicit in the calculation procedure. Failure to do this in busy laboratories or use of a design that does not permit such an inspection must surely be a major cause of unexpected or untrusted results, as well as of erroneous results that pass without detection.

4.9   Estimation of Potency by Interpolation from a Dose–Response Line Using an Automated System

In both macro- and microbiological assay, variation in observed responses (without definition as to whether the variation is of biological or physical origin) has long been regarded as a factor limiting the reliability of estimated potencies. For the purpose of the present discussion, setting aside considerations of specificity, efficiency in assay may be described qualitatively as maximal accuracy and precision for minimal practical effort. Progress toward this ideal may be attained by

(1)   defining and controlling practical conditions of assay so as to minimize random variation and bias,

(2)   employment of an appropriate experimental design (selected in accordance with principles such as are described in Chapter 8) so as to minimize the influence of random error on the final potency estimate.

In principle both these approaches to the problem are applicable to chemical and physicochemical methods, just as they are to biological methods of analysis.

Methods of analysis such as volumetric titrations and spectrophotometric measurements are in effect slope ratio assays. They are usually three point assays in which often the response to zero dose may be omitted and the "response" to standard calculated from a known constant such as an equivalent weight or an extinction coefficient. In contrast to biological methods, however, the basis of the quantitative relationship upon which the assay depends is in many cases known exactly. Often it can be described in terms of a simple chemical equation. In other cases, physical conditions can be controlled so as to give a highly reproducible relationship. In such circumstances there is no cause to use other than simple designs.

It is claimed by Kavanagh (1975a,b) that in the Autoturb® automated system for turbidimetric microbiological assay (which is described in Section 1.3), the treatment of tubes is so standardized and experimental conditions are so well controlled that reproducibility of responses is as good as obtained in many chemical methods. In this situation, a simple assay design may be used consisting of a standard dose–response curve from which sample potencies are interpolated at a single dose level.

In accordance with the recommendations of its designers, the Autoturb® system is used to prepare four dilutions in chilled inoculated nutrient medium from each test solution. Two dilutions consist of 0.1 ml of test solution diluted to 10 ml and the other two consist of 0.15 ml of test solution diluted to 10 ml. The test solutions are a series of doses of standard forming an arithmetic progression and one test solution for each sample.

The differing volumes of dose lead to differing volumes of inoculated broth used as diluent, that is, 9.90 and 9.85 ml, respectively. From a consideration of Eq. (4.1) it is clear that the difference of about 0.5% in initial concentration of living cells in the inoculum as well as perhaps other factors arising from the slight change in relative volume of the nutrient medium and test solution should influence the final concentration of cells. It is claimed that the Autoturb® system can detect this difference and that consequently two series of responses arising from different dose volumes should not be represented by a single dose–response curve.

In routine working, Kavanagh uses four standard curves and interpolates a single potency estimate for each sample from each of the four curves. This is illustrated by Example 16, the assay of erythromycin. The complete assay included 14 samples; however, results for only four of these are shown here. They were selected as a representative illustration of the reproducibility that was achieved in this assay.

Although this assay design does not include any formal statistical validity tests, any dissimilarity between sample and standard is likely to be revealed as a discrepancy in potency estimates derived from the different volume of dose. This would be highlighted by an unusually large relative standard deviation. The routine computer printout includes a figure for the standard deviation relative to the mean for the four potency estimates on each sample.

Example 16: Assay of a Series of Liquid Extracts Containing Erythromycin Using the Autoturb® Automated System for Turbidimetric Assay

Test organism: *Staphylococcus aureus*
Incubation: 4 hours at 35.0°C
Standard test solution potencies: 1.0, 2.0, 3.0, 4.0, 5.0 μg/ml
Sample test solutions: These were prepared so as to be close in potency to the midpoint of the standard curve.
Observations and computer interpolated potencies of sample test solutions are summarized in Table 4.8. The form of the dose–response relationship is shown in the two curves of Fig. 4.7. Although the computer interpolates from the two individual curves for each dose volume, it is impracticable to show these separately as duplicate responses are so close.

Table 4.8 Data of Example 16

Volume of dose:	Response, (absorbance × 100)				Interpolated potency (µg/ml)				Mean potency	RSD[a]
	0.10 ml		0.15 ml		0.10 ml		0.15 ml			
Zero control	26.24	26.20	26.07	26.07						
Standard, µg/ml										
1	21.90	21.83	18.94	18.80						
2	17.08	17.11	13.75	13.60						
3	13.93	13.77	10.81	10.78						
4	11.90	11.94	9.11	9.14						
5	10.33	10.27	7.85	7.84						
Sample number										
1	11.85	11.78	9.07	9.07	4.024	4.086	4.032	4.046	4.047	0.69
2	13.87	13.79	10.78	10.74	3.030	2.994	3.015	3.025	3.016	0.53
3	15.18	15.06	12.04	12.02	2.578	2.588	2.553	2.532	2.563	0.98
13	13.90	13.89	10.90	10.85	3.016	2.962	2.966	2.976	2.980	0.84

[a] This column gives the relative standard deviation (standard deviation relative to the mean) of the four potency estimates on each sample. This figure is obtained as part of the routine computer printout.

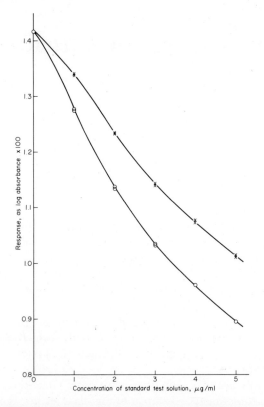

Fig. 4.7. Graphical representation of the dose–response relationship in the automated turbidimetric assay of erythromycin using the Autoturb[R] system. As duplicate responses to the same dose and volume of test solution lie so close together, a composite curve is drawn for the 0.10 ml dose volume responses and a separate composite curve is drawn for the 0.15 ml dose volume responses. The four potency estimates for each sample, however, are calculated by the computer by interpolation from the four individual curves. ×, individual responses to 0.10 ml doses; ○, individual responses to 0.15 ml doses.

References

Berkson, J. (1944). *J. Amer. Statist. Assoc.* **39**, 357.
Bliss, C. I. (1934a). *Science* **79**, 38.
Bliss, C. I. (1934b). *Science* **79**, 409.
Fisher, R. A., and Yates, F. (1963). "Statistical Tables for Use in Biological, Agricultural and Medical Research." Longman, London.
Gaddum, J. H. (1933). Medical Research Council, Spec. Rep. Ser. No. 183.
Garrett, E. R., and Wright, O. K. (1967). *J. Pharm. Sci.* **56**, 1576.
Hemmingsen, A. M. (1933). *Quart. J. Pharm. Pharmacol.* **6**, 39, 187.
Kavanagh, F. W. (1968). *Appl. Microbiol.* **16**, 777.
Kavanagh, F. W. (ed.) (1972). "Analytical Microbiology," Vol. II. Academic Press, New York.
Kavanagh, F. W. (1975a). *J. Pharm. Sci.* **64**, 844.
Kavanagh, F. W. (1975b). Personal communication.

CHAPTER 5

ASSAY OF MIXTURES OF ANTIBIOTICS

5.1 Occurrence of Mixtures and the Nature of the Problem

Mixtures of two or more antibiotics are encountered for one of two reasons. Either they are produced naturally during fermentation or they are components of a pharmaceutical formulation.

Naturally occurring mixtures often include a series of related substances as in the case of the penicillins, bacitracins, and neomycins.

Developments in fermentation and the use of precursors have resulted in production of penicillins consisting predominantly of the required species. The ability to produce the required species is reflected in the International Pharmacopeia monograph for benzylpenicillin sodium. Not only is it required to contain not less than 95% of total penicillins, it must also contain not less than 90% of benzylpenicillin, both calculated as the sodium salt. This has been made possible by the use of phenylacetic acid as a precursor; thus the species with the benzyl side chain is preferentially formed. Similarly, in the production of phenoxymethylpenicillin (penicillin V) phenoxyacetic acid is used as precursor.

Other naturally occurring mixtures may contain unrelated or at least not closely related components. Examples of this class include tyrothricin (components gramicidin and tyrocidine) and the natural cephalosporins (components C, N, and P).

Combinations of antibiotics in pharmaceutical formulations may be of completely unrelated substances such as streptomycin and penicillin, related substances such as streptomycin and dihydrostreptomycin or merely different salts of the same active substances.

Having outlined the types of mixtures that may be encountered, it is necessary to consider first the nature of the problem. In which cases is it necessary to estimate the separate active components? In which cases is a measure of total activity both meaningful and useful?

A few examples will serve to indicate the principles involved in trying to resolve these problems. A satisfactory solution is not always possible.

Neomycin presents a particularly interesting and difficult problem. Its major component is neomycin B, but it contains a significant proportion of neomycin C (perhaps 10 to 30%), which is less active than B against a variety of organisms. A third component is neomycin A, more commonly known as neamine. This is a decomposition product of both B and C and occurs to the extent of perhaps 1 to 5%. Neomycins B and C are glycosides of the organic base neamine. They are of identical molecular weight, differ only in the configuration of an aminoethyl group, and are not easily separable by normal analytical procedures. The differing ratios of components B and C in commercial samples of neomycin result in real difficulties in standardization of the product.

Sokolski *et al.* (1964) discuss the different behavior of components B and C according to composition of assay medium, especially its salt content. By choice of a medium of low ionic content and the right test organism, an assay system can be devised in which components B and C respond equally. However, unless it can be shown that neomycins B and C are equipotent *in vivo* against a variety of infecting organisms, then such an approach to assaying the neomycin complex disguises rather than solves the problem.

Lightbown (1961) refers to the difficulty in obtaining agreement between successive assays in either the same or different laboratories, as well as the difficulty of getting statistically valid assays with parallel response lines. He cites the reason for this as being the varying effect on the different components by changes in pH, E_h, composition of medium, temperature, and age of the test organism.

Hewitt *et al.* (1959), using diffusion assays, were able to obtain reasonably consistent potency estimates for the isolated components B and C. Sample and standard dose–response lines were parallel in assays using either *Staphylococcus aureus* or *Sarcina lutea*. The estimated potencies of these isolated components varied, of course, according to test organism.

Since the two components B and C are so closely related and are of identical molecular weight, it seemed reasonable that their diffusion characteristics might also be very similar thus resulting in parallel responses regardless of test organism.

A further paper by Lightbown (1969) shows that the nonparallel dose–response lines for neomycins B and C were obtained when using a very wide dose range, 20–1000 μg/ml. He suggests, however, that in an assay with the normal two- or fourfold dose range, deviation from parallelism might appear statistically nonsignificant.

A study of commercially available neomycins from many countries was made for the World Health Organization (1970). It revealed that the then current International Reference Preparation was unrepresentative of the commercially available material with regard to proportions of components B and C. This was in contrast to the state of affairs when the standard was first established in 1958.

Wilson *et al.* (1973) showed by gas liquid chromatography that in neomycin products available on the Canadian market, the proportion of neomycin C varied from 2.5 to 31%.

To summarize, one example has been given of a mixture of varying proportions of closely related antibiotic substances. The accepted methods of assay depend on measurement of total antibiotic activity. The limitations of this convention are recognized. No true comparison of sample with a reference standard is possible unless both preparations contain active ingredients in the same relative proportions. Such information is almost invariably not available.

The general case was stated by Lightbown (1961):

> For heterogeneous materials controlled biologically and assayed against a heterogeneous standard, it must be recognized that there is no true potency for any particular sample. A sample will have a family of potencies depending on the conditions of assay. These may be distributed about a mode, but the modal value has no intrinsic superiority over any individual value.

The assessment of total penicillin activity no longer presents such problems, because nowadays the proportion of the major constituent varies only slightly from batch to batch.

Statistically valid assays comparing the potencies of streptomycin and dihydrostreptomycin are possible although the potency ratio may vary according to the test organism (Miles, 1952). In the assessment of potency of mixtures of the two substances in pharmaceutical formulations, a mixed standard has been employed. This is of some value in the measurement of total activity in stability studies. In production quality control however, the logical procedure is (a) to check the potency of the individual components before mixing and (b) to check the proportion in the blend by specific methods. This is conveniently achieved by chemical methods.

Procaine and other relatively insoluble salts of benzylpenicillin may be assayed by dissolving in a small quantity of alcohol or dimethylformamide and then diluting to assay level with the same buffer as the standard. The standard used is the normal penicillin standard, i.e., the sodium or potassium salt. The assay is a straightforward comparison of the same ionic species (benzylpenicillin) in the presence of an excess of the same buffer.

Assuming that sodium benzylpenicillin of 100% purity has a potency of 1670 IU/mg, the theoretical potency of the procaine salt may be calculated from the ratio of molecular weights:

$$(1670 \times 356.4)/588.7 = 1011 \quad IU/mg.$$

Mixtures of different salts of penicillin such as occur in pharmaceutical formulations may also be assayed against the normal working standard. A typical case is "Fortified Injection of Procaine Penicillin" of the British Pharmacopeia. This contains 300,000 units in one dose, in the form of the

procaine salt, and 100,000 units as the sodium or potassium salt. A determination of total penicillins (whether by microbiological assay or chemical
assay) is of value in the examination of old samples as a check on deterioration.
It is virtually valueless in routine testing of production batches, however, as
it is not an effective means of checking the composition of the mixture. The
logical procedure is to check the potency of the input ingredients. A colorimetric determination of the procaine base content of the compounded form
provides a simple and accurate check that the proportion of the procaine
salt is within specification limits. If the procaine salt is in the correct proportion, it may be inferred that the proportion of sodium or potassium salt
is also very likely to be correct.

The reference to specification limits is important. Realistic upper and
lower limits need to be defined in a company manufacturing specification.
The B.P. monograph is not intended as a manufacturing specification and is
quite unsuitable for that purpose. See also Section 9.5.

Unlike salts, esters cannot be assayed directly in terms of the parent
antibiotic. Chloramphenicol palmitate and chloramphenicol sodium succinate, for example, do not posess *in vitro* activity. They are usually assayed
by means of their ultraviolet absorption spectrum.

Stating the general case, any ester that did display *in vitro* activity would
very probably have different diffusion characteristics from the parent antibiotic, and so for a plate assay the standard would have to be of the same
ester.

5.2 General Techniques

When it is necessary to determine the separate components of a mixture,
this may be achieved by one or more of the basic methods; that is, (a) separation by standard analytical techniques followed by assay of the separate
components, (b) selective destruction of one component followed by assay
of that remaining, (c) selective determination by use of a test organism
sensitive to only one of the components under the defined conditions of test.

(a) *Separation methods.* Depending on relative solubilities or partition
coefficients, solvent extraction may be a convenient way of separating antibiotics from one another or from excipients. Weiss and co-workers (1957,
1959, 1967) measured the solubilities of more than 120 antibiotics in over 20
different solvents. Their tabulated findings are quoted by Kavanagh (1963,
1972) in Volumes I and II of "Analytical Microbiology." Kirshbaum (1972)
describes the use of a cation exchange resin for the in situ removal of basic
antibiotics such as streptomycin when assaying novobiocin by the agar

diffusion method. A suspension of finely powdered, sterilized resin is added to the seed agar at the time of inoculation. Streptomycin is held by the resin as it begins to move outward from the reservoir.

Goodall and Levy (1947) described a paper chromatographic method for separation of the various penicillins in broth cultures. The developed paper strip chromatograms of samples and standard were laid on large rectangular plates of inoculated medium. The antibiotic diffused from the paper into the agar gel so that on incubation, inhibition zones were produced. These were identified by their positions and quantitatively estimated by comparison of zone sizes. This technique, known as the bioautograph, has been used by Stephens and Grainger (1955) for the qualitative examination of penicillin V fermentation liquors. It was also used by Winsten and Eigen (1949) for the qualitative examination of naturally occurring mixtures of vitamin B_{12} and its analogs. In this case exhibition zones were produced on a synthetic medium inoculated with *Escherichia coli*.

During the early stages of development of an antibiotic known by the code name of E129, Hewitt *et al.* (1957) used a bioautograph technique that was quantitative in that it determined the relative proportions (activities) of two components. Areas of the zones were related to the logarithm of the relative doses.

When pure reference materials for these two components became available, the method was put on a truly quantitative basis by Bessel *et al.* (1958).

The quantitative aspects of these methods for E129 are shown in detail in Sections 5.3 and 5.4.

The United States Code of Federal Regulations describes a paper chromatographic method for the examination of the gentamycin complex. The positions of the three separated components C_1, C_{1a}, and C_2 are located by the use of a duplicate chromatogram, which is sprayed with ninhydrin reagent. The appropriate zones are cut out of the unsprayed paper and extracted with buffer solution; the resulting extracts are then assayed by the normal agar diffusion assay.

Wagman *et al.* (1968) also describe a quantitative bioautograph technique for the gentamycin complex.

Lightbown and Rossi (1965) described bioautograph procedures following separation by electrophoresis. The behavior of 24 antibiotics was studied. Quantitative determinations were based on zone diameter. Precision of the method appears to be about the same as for the normal plate assay. The technique was also used to overcome the effects of serum binding in estimating penicillin in blood. Separation, it was claimed, had advantages over the technique of adding serum to standard, as the latter was subject to bias according to the varying binding properties of the sera in sample and standard preparations.

(b)   *Selective destruction methods.*   The principles of selective destruction methods are simple. In a mixture of two antibiotics, designated A and B, the assay of A is accomplished by first decomposing B, which would otherwise interfere. The treatment should either completely inactivate B or at least reduce the concentration of B remaining to a level such that it does not interfere with the determination of A. In addition to being inactive themselves, decomposition products of B should not modify the activity of A in any way. Reagents and treatment used for inactivation of B should preferably have no effect on A. However, compensation for small effects may be achieved by application of exactly the same treatment to the reference standards.

Before applying an inactivation procedure routinely, the analyst should confirm by simple tests in his own laboratory that results are in conformity with these principles.

Only a few examples will be given here. For others the reader may consult the literature to obtain detailed procedures for particular mixtures.

In the estimation of yields of cephalosporin C, in fermentation broths, treatment with dilute hydrochloric acid has been used to destroy cephalosporin N. Samples were adjusted to pH 2 and held at room temperature ($\sim 20°$) for 2 hours prior to neutralization and dilution with buffer to assay for cephalosporin C content.

In the assay of mixtures of either novobiocin or erythromycin with penicillin, the latter may be inactivated by penicillinase.

Kaplan *et al.* (1965) refer to the use of a cupric–morpholine complex, which rapidly inactivates tetracycline in methanol or aqueous solutions. This permits the assay of kanamycin when mixed with tetracycline.

(c)   *Selective determination methods.*   These range from the specific determination of one of the antibiotics by suitable choice of test organism, to the cases where no organism can be found that responds to one component yet fails completely to respond to the other.

Clearly the former is much more satisfactory. Examples include the assay of bacitracin in the presence of neomycin by the plate method using *Micrococcous flavus* and the assay of neomycin in the presence of bacitracin using *Micrococcus albus*.

Kavanagh (1963, 1972) reproduces tables of interference thresholds devised by Arret *et al.* (1957, 1968). For the assay of certain antibiotics A, these tables show the limiting tolerable concentrations of potentially interfering antibiotics B. These are a very valuable guide but should of course be checked by experiments in the analyst's own laboratory, as varying assay conditions may well lead to significantly different interference thresholds. Arret *et al.* considered there was interference if the result lay outside the limits 90 to 110% of the potency of the test solution containing only the single antibiotic.

It should be noted that a bacteriostatic antibiotic may reduce the effectiveness of a bactericidal antibiotic, thus causing a drop in determined potency. These rather wide tolerences set by Arret were acceptable in assays conducted for screening purposes.

However, when a more reliable estimate of potency is needed, stricter criteria should be applied; interference must be considered not only at one dose level of the antibiotic to be assayed, but at every level of the standard curve, or at least at the two extreme levels. Response lines for A alone and A plus B should be evaluated for departure from parallelism.

It should not be construed, however, that interference is necessarily accompanied by a nonparallel response. Considering the cases in which it is not possible to find entirely selective test organisms, a differential assay may be devised. Such an assay is dependent for its validity on the fact that the response lines have been shown to be parallel. An example is given in Section 5.5, the differential assay for neomycins B and C. This of course is a naturally occurring mixture and is not one of the problems that Arret studied. Nevertheless, the principles involved in the neomycin differential assay are equally applicable to compounded mixtures.

5.3   Comparative Bioautographs

The bioautograph technique has been used as a qualitative and implied semiquantitative method for the evaluation of naturally occurring mixtures of antibiotics and vitamins. For example, other penicillins in penicillin V and related substances in crude samples of cyanocobalamin. Such was the case during development stages of the antibiotic known by the code name of E129. It was known to contain two components A and G and was assayed for total activity against an arbitrary standard of unknown exact composition using *S. lutea* by the normal diffusion assay.

The ratio of activities of A to G was assessed visually on the basis of a bioautograph method.

The method in outline was as follows: Doses of samples as well as the arbitrary standard containing both components were applied to a sheet of chromatography paper using pipettes that delivered approximately 20 μl. The same pipette was used for each spot on one piece of paper. Chromatograms were developed in a solvent system of water, methanol, benzene, and cyclohexane. After drying, the papers were applied to large plates seeded with *S. lutea*. After allowing time for diffusion of the antibiotic into the agar (about 15 min) the papers were removed and the plates incubated overnight. Clear zones marked the position of the separated components A and G. Methylene blue prints were obtained by the method described by Stephens and Grainger (1955) for the bioautograph of penicillin V. Plates were flooded with a solution of methylene blue and the dye was allowed to diffuse into the agar for a few minutes. The dye solution was poured off, the plates washed in running water, and then surplus water removed from the agar surface with blotting paper. On applying a sheet of chromatography paper to the agar surface, dye diffused from the clear zones onto the paper to give a blue print of the areas of inhibition. In areas where growth had occurred, the dye was absorbed by the cells and little was free to diffuse on to the paper.

Inconsistencies between the visually assessed proportions of A to G in the various stages of extraction and purification led Hewitt *et al.* (1957) to attempt to put this method on a more quantitative basis.

Bioautographs of a single preparation at a series of doses were prepared. Methylene blue prints were dried and the zone areas measured. Zones were cut out with scissors and weighed or the outline could be traced with a planimeter, which measured area without destroying the print.

Results of a typical test are given in Example 17 and the appearance of the corresponding print is illustrated in Fig. 5.1. This print makes clear the fallacy of attempting to estimate proportions of components by purely visual inspection.

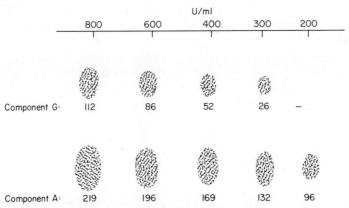

Fig. 5.1. The comparative bioautograph of antibiotic E129. This illustrates the appearance of the zones on the methylene blue print obtained from a bioautograph. The observations relate to Example 17. A single sample of the antibiotic was used at dose levels ranging from 200 to 800 units of total activity per milliliter. The areas of the inhibition zones are indicated by the numbers under each zone.

No estimate of the ratio of the two components A and G can be made by visual inspection of the print. It was calculated from these results that component G contributes about 23% of the total A + G activity versus *S. lutea.* (Hewitt *et al.,* 1957.)

Example 17:   Comparative Bioautograph of Antibiotic E129 to Determine Proportions of Components A and G

Test organism: *Sarcina lutea*
Sample: Provisional reference standard
Responses: Inhibition zone areas measured as weight of paper cut from methylene blue print
Observations: See Table 5.1

Table 5.1

Dose, U/ml:	800	600	400	300	200
Responses (mg)					
Component A	219	196	169	132	96
Component G	112	86	52	26	—

Observations plotted in Fig. 5.2 with doses on a logarithmic scale and responses on an arithmetic scale give two straight parallel lines. Proportions of A to G are estimated from the intercepts of any convenient vertical line on the two response lines. In Fig. 5.2 a vertical line is drawn at response 104 mg.

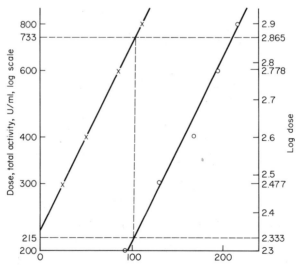

Fig. 5.2.   Graphical representation of the comparative bioautograph assay of the antibiotic E129, Example 17, using *S. lutea* ATCC 9341. Responses (area of inhibition zones) when plotted against log dose gave straight lines. The response lines for components A and G are parallel, thus making possible a comparison of their activities. It was calculated that component G contributes about 23% of the total A + G activity of this sample versus *S. lutea*. Component G, × ; Component A, ○. (Hewitt *et al.*, 1957.)

Potencies corresponding to the intercepts are obtained by:

response line A: log dose = 2.333, dose = 215 U/ml
response line G: log dose = 2.865, dose = 733 U/ml

That is, the G activity in 733 units of total activity is equal to the A activity in 215 units of total activity. The G : A activity ratio in the sample is therefore 215:733. The result as G activity as percentage of total activity is

$$(100 \times 215)/(733 + 215) = 22.7\%$$

5.4 Quantitative Bioautographs

These will be illustrated by the development of the E129 bioautograph method.

When purified samples of components A and G became available for use as standards, the assay was put on a truly quantitative basis by Bessell *et al.* (1958). By this time a third major component, designated factor B, had been isolated. This was not particularly active against *S. lutea* in itself but had a synergistic effect, enhancing the activity of both factors A and G. Factor B was separable from A and G by the chromatograph system previously described, but trailing of B (which moved faster than A) tended to cause interaction with the zone for A, resulting in elongation of the A zone. This effect was overcome by the inclusion of a subinhibitory concentration of factor B in the assay medium, so that the maximum synergistic effect was obtained on all zones regardless of trailing of the B spot.

In devising an assay design, the size of the chromatography tanks, papers, and agar plates imposed the restriction that not more than five doses could be applied to each paper. In these circumstances the most efficient utilization of resources was by employment of a design including three dose levels of standard and two of sample.

Normally four plates were used for each assay and the positions of test solutions were in accordance with four different randomized designs to minimize positional bias.

In Example 18 this calculation procedure is illustrated. Only observations for factor G are shown here, since the calculation method for factor A is identical.

Example 18: Quantitative Bioautograph of Antibiotic E129 to Determine Amounts of Components A and G

Test organism: *S. lutea* ATCC 9341
Standards: Mixed solution prepared from purified components A and G

Sample: A solution of E129 diluted 5 ml → 100 ml to test solution level (high dose)
Test solution concentrations:

	Standard (μg/ml)			Sample μg/ml (nominal)	
factor A:	12	6	3	9	4.5
factor G:	24	12	6	18	9

Observations: See Table 5.2; factor G only, inhibition zone weight of paper (mg)

Table 5.2

| Dose, μg/ml | Standard | | | Sample | |
	24	12	6	18	9
Responses	156	96	39	129	69
	159	106	37	136	71
	165	114	57	143	85
	162	109	48	137	69
Totals	642	425	181	545	294
Means	160.50	106.25	45.25	136.25	74.50

Difference due to dose:

$$E = \tfrac{1}{5}[(2S_3 + T_2) - (2S_1 + T_1)] = \tfrac{1}{5}[457.25 - 165.00] = 58.45$$

Difference due to sample:

$$F = \tfrac{1}{2}(T_1 + T_2) - \tfrac{1}{3}(S_1 + S_2 + S_3) = \tfrac{1}{2}(210.75) - \tfrac{1}{3}(312.00) = 1.38$$

Slope:

$$b = E/I = 58.45/0.3010 = 194.19$$

As means of standard dose and nominal sample dose are unequal, the ratio of estimated to nominal potency for the sample test solution is obtained via Eq. (2.5b) using

$$\bar{x}_S = \log 12 = 1.0792$$
$$\bar{x}_T = \tfrac{1}{2}(\log 18 + \log 9) = 1.1048$$
$$M = F/b = 1.38/194.19 = 0.0071$$

Thus

$$M' = 0.0071 + 1.0792 - 1.1048 = -0.0185 = \bar{1}.9815$$

R', the ratio of actual sample test solution potency to its nominal potency, is obtained as antilog of M', i.e.,

$$R' = \text{antilog } \bar{1}.9815 = 0.9583$$

so that sample potency is given by

$$\frac{18 \times 0.9583 \times 100}{5} = 345 \quad \mu g/ml \text{ of component G}$$

It will be noted that in the calculation of E, the slope due to response to standard is weighted by a factor of 2 as it is measured over two dose intervals. Thus the sum within the square brackets corresponds to a difference over five dose intervals.

5.5   Differential Assays

In an earlier part of this chapter, reference was made to the problems arising from differing responses by different test organisms to the individual components in a mixture of antibiotic substances.

Although in some respects a problem, it is a phenomenon of some value. Differing relative potencies between two substances, according to the test organism used, alerts the investigator to the possibility that there are qualitative differences between the two. It was the differential responses of *Staphylococcus aureus* and *Bacillus subtilis* to various samples of penicillin that led to the recognition of penicillin as a family of related substances by Schmidt *et al.* (1945).

If the differential between responses to different organisms is great enough, it may be possible to use it as a means of estimating the proportion of the components in the samples.

Such a system was used by Hewitt *et al.* (1959) in the examination of neomycin samples to determine relative proportions of components B and C. They developed a differential assay based on potencies determined in their own laboratory for the isolated components B and C against *S. aureus* and *S. lutea*.

The basic assumptions for the assay were:

(1)   The potency of the components B and C would remain constant provided that there was no change in the assay conditions and the reference standard.

(2)   Active components other than neomycins B and C were present in small proportions and had only a negligible influence on the potency as determined by either test organism.

(3)   Activities of the two components were purely additive, that is to say, there were neither synergistic nor antagonistic effects.

(It had been demonstrated that the quantities of neamine normally present had an insignificant effect).

Assay variables were minimized by using the same test solutions for assays with both organisms. Thus, the only factors affecting potency ratio were the true difference in response according to test organism and the random error of the response.

Potencies of the pure components used to establish this differential assay had been determined in comparison with the company working standard:

neomycin B sulfate versus *S. aureus*   850 U/mg
                        versus *S. lutea*   950 U/mg
neomycin C sulfate versus *S. aureus*   410 U/mg
                        versus *S. lutea*   15 U/mg

Proportions of the two components in a mixture could be estimated by substituting in the simultaneous equations

$$850b + 410c = \text{potency versus } S.\ aureus$$
$$950b + 15c = \text{potency versus } S.\ lutea$$

where b and c correspond to the proportions of the two components neomycins B and C, respectively, in the sample.

The percentage of the two components may be calculated as

$$\%B = 100b/(b + c)$$
$$\%C = 100c/(b + c)$$

However it was more convenient to construct a graph relating potency ratio to %B and %C. This was done by assuming a series of proportions of B to C, calculating the corresponding potency ratios and plotting these graphically. (Fig. 5.3)

For example, assuming 70% B and 30% C:

(1) potency versus *S. aureus*   70% B at 850 U/mg = 595 U/mg
                                       30% C at 410 U/mg = 123 U/mg
                                       ─────────────
                                                         718 U/mg

(2) potency versus *S. lutea*   70% B at 950 U/mg = 665 U/mg
                                       30% C at 15 U/mg =   5 U/mg
                                       ─────────────
                                                         670 U/mg

The potency ratio is $718/670 = 1.07$.

The precision of differential assays in general is limited by the degree of contrast in activity of the two components toward the two test organisms. In the case of neomycin the contrast is moderately good. However, due to the inherent low precision of the neomycin diffusion assay the degree of replication of observations needed to obtain a meaningful result is very substantial. Suppose that ratio of potency estimates between the two assays is to be determined with confidence limits of $\pm 3\%$ (P = 0.95). From Fig. 5.3

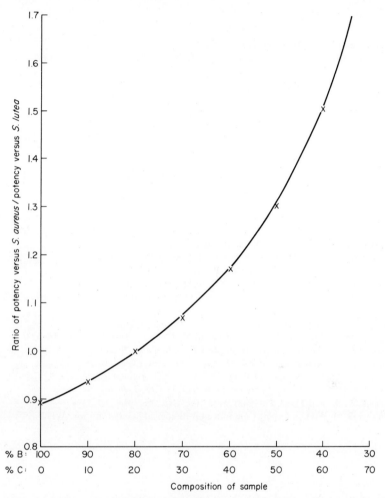

Fig. 5.3. Graphical representation of the different values of the ratio potency versus *S. aureus*: potency versus *S. lutea* for samples of the neomycin complex differing in proportions of components B and C. The curve was used to determine proportions of components B and C from the ratio of potencies determined using these two test organisms. It applies only to potency ratios determined under defined conditions of test and against a particular reference standard. (Hewitt *et al.*, 1959.)

it may be seen that a potency ratio of 1.30 (limits 1.26–1.34) corresponds to a neomycin B content of 50% (limits 47.5–52.5%), whereas a potency ratio of 0.95 (limits 0.92–0.98) corresponds to a neomycin B content of 88% (limits 83–93.5%). Thus the range of the confidence limits increases at higher percentages of neomycin B.

References

Arret, B., and Eckert, J. (1968). *J. Pharm. Sci.* **57**, 871.

Arret, B., Woodard, M. R., Wintermere, D. M., and Kirshbaum, A. (1957). *Antibiot. Chemother.* **7**, 545.

Bessell, C. J., Fantes, K. H., Hewitt, W., Muggleton, P. W., and Tootill, J. P. R. (1958). *Biochem. J.* **67**, 24P.

Goodall, R. R., and Levy, A. A. (1947). *Analyst* **72**, 274.

Hewitt, W., Buckingham, W., and Chesterman, D. (1957). Unpublished data.

Hewitt, W., Harris, D. F., Smith, W. C. J., and Stoddart, G. A. (1959). Unpublished data.

Kaplan, M. A., Lannon, J. H., and Backwalter, F. H. (1965). *J. Pharm. Sci.* **54**, 163.

Kavanagh, F. W. (1963 and 1972). "Analytical Microbiology," Vols. I and II. Academic Press, New York.

Kirshbaum, A. (1972). *In* "Analytical Microbiology" (F. W. Kavanagh, ed.), Vol. II, pp. 315–319. Academic Press, New York.

Lightbown, J. W. (1961). *Analyst* **86**, 216.

Lightbown, J. W. (1969). *In* "The Control of Chemotherapy, Report on a Symposium" (P. J. Watt, ed.). Livingstone, Edinburgh and London.

Lightbown, J. W., and De Rossi, P. (1965). *Analyst* **90**, 89.

Miles, A. A. (1952). World Health Organization Monogr. Ser. No. 10, p. 131.

Schmidt, W. H., Ward, G. E., and Coghill, R. D. (1945). *J. Bacteriol.* **49**, 411.

Sokolski, W. T., Chidester, C. G., Carpenter, O. S., and Kaneshiro, W. M. (1964). *J. Pharm. Sci.* **53**, 826.

Stephens, J., and Grainger, A. (1955). *J. Pharm. Pharmacol.* **7**, 702.

Wagman, G. H., Oden, E. M., and Weinstein, M. J. (1968). *Appl. Microbiol.* **16**, 624.

Weiss, P. J., and Andrew, M. L. (1959). *Antibiot. Chemother.* **9**, 277.

Weiss, P. J., and Marsh, J. R. (1967). *J. Assoc. Off. Anal. Chem.* **50**, 457.

Weiss, P. J., Andrew, M. L., and Wright, W. W. (1957). *Antibiot. Chemother.* **7**, 374.

Wilson, W. L., Belec, G., and Hughes, D. W. (1973). *Can. J. Pharm. Sci.* **8**, 48,

Winsten, W. A., and Eigen, E. (1948). *Proc. Soc. Exp. Biol. Med.* **67**, 513.

W.H.O. (1970). W.H.O. Expert Committee on Biological Standardisation, WHO/BS/70.1001.

CHAPTER 6

EVALUATION OF PARALLEL LINE ASSAYS

6.1 Introduction

In Chapters 2–4 the raw data from various experiments have been processed by appropriate means to give an estimate of the potency of each sample tested. The same raw data can be processed further to yield more information, including estimates of the precision of the determined potency as well as of the validity of the assay. For example, for log dose–response assays it is possible to assess whether deviations from parallelism between sample and standard responses are real (an invalid assay) or whether they may be attributed to random variation.

It has been said that the statistical analysis is an integral part of the calculation of results of a biological assay. Also having made a lot of practical effort in obtaining the raw data it is surely worthwhile to expend a little more effort on statistical calculation to obtain a more meaningful result.

While this is undoubtedly true for biological assays using animals, it is certainly not always true for microbiological assays. Unless a computer were available, the additional work in applying statistical analysis routinely to microbiological assays would represent an unjustifiable and intolerable burden on a busy laboratory.

It is to emphasize this point that statistical analysis is treated here in separate chapters.

For any analytical method, be it chemical, physical, or biological, both the analyst and the recipient of the quantitative reports are concerned to know their reliability.

Is the method free from bias? How reproducible are the results? What is the probable range about the reported result in which the true result will be found?

To answer these questions in the case of physical or chemical analysis, the analyst will base his conclusions on his experience of the method's reproducibility, possibly having determined the standard deviation of the method in the hands of one or more operators.

He will possibly have examined samples of known composition to compare known concentration with concentration found. He may also have compared the method with an alternative method using the same sample.

These procedures are equally applicable to microbiological assays for which reproducability (manual assays) may be not much less than for some chemical methods. Reproducability of some automated microbiological methods is high. The reader may wonder then why statistical analysis is applied to microbiological assays.

In the case of the diffusion method, statistical evaluation reveals the following:

(1)   It isolates and measures variation of zone size arising from certain identifiable causes such as differences between plates and differences between test solutions.

(2)   It leads to a figure for variance of zone size—excluding the influences described in (1)—which cannot be further broken down and distinguished from random error. This is designated "residual error," $V(y)$. Important sources contributing to this residual error are clarity of the zone edge, ability of the operator to do precise work (eyesight), and precision of the measuring system. Other contributing sources may be some of those unwanted influences described in Section 2.2, which remain despite practical techniques and experimental design employed to minimize them.

(3)   Using the variance of the zone size together with the estimate of logarithm of sample potency and data on assay design, confidence limits for the estimated potency may be calculated.

The practical value of statistical analysis is therefore threefold:

(1)   An occasional check contributes to the laboratory supervisor's knowledge of the capability, validity, and reliability of the method as carried out in his laboratory.

(2)   In certain cases, such as the establishment of a national or a company reference standard or to conform with official requirements, it is necessary to know the confidence limits for the estimated potency.

(3)   A knowledge of the general principles of statistical evaluation leads to an understanding of the capabilities and the limitations of the various designs that have been proposed for use in biological assays.

Even though there are circumstances where the calculation of confidence limits and applications of validity checks is demanded by a regulatory authority, there seems little doubt that this third point is the major value of an understanding of statistical evaluation in relation to *micro*biological assay. This will become clear to the reader when he has worked through the examples of this chapter and then studied the relevant parts of Chapter 8.

If practical conditions of assay are well controlled and standardized, and confidence limits have been determined for the assay of a particular substance by a certain design two or three times, then it can be reasonably assumed that confidence limits will vary only little in repeated assays by that same design. Knowing the principles of assay design, it is then a simple matter to forecast how change of design would influence width of confidence limits. There is no necessity actually to do the evaluation following a change of design.

The many examples given in this chapter serve to illustrate the principles that are discussed in Chapter 8. This indeed is their main purpose and in fact the sole purpose of some of the examples. Thus, having seen the limitations of the standard curve method in Example 7, the low precision assay of bacitracin, the reader might find that such an economical assay would be acceptable for some purposes. Knowing its limitations, there would be little to be gained by repeating the mathematical exercise that is shown here as Example 7e.

The limitations as well as the usefulness of these mathematical techniques should always be borne in mind. No more information can be obtained from them than is put into them. Gross errors in weighing or dilution could pass completely undetected unless the experimental design included appropriate checks. Such checks are not included in routine work. However the 12 × 12 Latin square designs using very large plates as described by Lees and Tootill (1955) do include duplicate weighings of both sample and standard, so that discrepancies between the duplicate weighings or dilutions would be revealed in the statistical examination. Such errors should of course be rare. They could become significant, however, when quantity of material is very limited and weighings are, for example, less than 10 mg. An error of 0.1 mg would then represent a bias of 1% or more and would defeat the whole object of these high precision assays.

An error affecting both duplicates equally would not be detected. Bias due to an interfering active substance present in the sample but absent from the standard may also pass undetected. On the other hand, it might be revealed if it caused a nonparallel sample response or it might be revealed by a differential response in two assays employing different test organisms.

In short, the use of statistical techniques that produce fiducial limits for potency reveals only part of the story. The analyst is in no way absolved from the need to use the common sense and logic that he regularly applies in the assessment of chemical or physical methods. The examples in this chapter that are designated "e" (evaluation) are extensions of those examples with the same number in earlier chapters. It is sometimes necessary to refer back to the initial part of the calculation.

In statistical evaluation it is often necessary to retain a large number of significant figures at intermediate stages, where they may be useful for checking purposes. Sometimes the difference between two very large numbers is small. Realistic rounding off is done at the end.

6.2 Basic Assumptions for the Statistical Evaluation of Parallel Line Assays

The methods of statistical evaluation for agar diffusion and other parallel line assays are dependent on the following assumptions:

(1)   The response or function of response used in the calculation is a rectilinear function of the logarithm of the dose. It is stressed that this is a requirement for validity of the statistical technique and not for the fundamental validity of the assay itself. Empirical linearizing techniques may be employed so that experimental data can be made to fit the statistical requirements. The author has done this on occasion, using the square of zone diameters in place of zone diameters themselves. In this way statistical requirements for validity were met but there was no perceptible change in the calculated confidence limits.

(2)   The log dose–response lines for sample and standard are parallel.

(3)   Test solutions are accurately prepared so that error in doses is negligible.

(4)   Replicate responses to the same test solution vary in a random manner and are normally distributed.

(5)   The variance of the response does not change according to dose level. However, Emmens (1948) states that this is not so critical provided that the assay is balanced and mean squares for sample and standard are not much different.

Many assay designs permit checks that there are not serious departures from assumptions (1) and (2).

As was described in Section 2.2, many factors other than dose have an effect on zone size. In a well planned and executed experiment these unwanted effects are minimized by good technique. The residual effects are partly balanced out by randomization, which tends to reduce systematic bias and thus helps to ensure that responses to an individual test solution are normally distributed as required by assumption (4).

It can be demonstrated by determining standard deviation of response separately for each dose level in an assay that variance of the response does not usually change significantly with differing dose level within the assay.

6.3   Evaluation of a Standard Log Dose–Response Curve

Example 1 illustrates a standard response line for varying doses of streptomycin on plates inoculated with *Bacillus subtilis*. As was explained in Chapter 2, this was done purely to establish what were suitable conditions for future assays. As no samples were assayed on this occasion it is only possible to take the evaluation as far as estimation of variance and standard deviation of zone diameter.

This is useful in that it gives an indication of the capability of this streptomycin assay method (see Chapter 8). Here it serves as an introduction to the principles of evaluation. The data are treated first (Example 1e(a)) in the general manner that would be applicable if doses were randomly distributed between the plates and if dose levels did not form a geometrical progression.

Example 1e(a): Estimation of Variance and Standard Deviation of Zone Diameter from a Standard Curve for Streptomycin (*Bacillus subtilis,* Agar Diffusion Assay Method)

In this calculation in order to simplify the arithmetic, all the responses of Example 1 are modified by first subtracting 10 from each and then multiplying by 10. Thus, 19.3 becomes 93. As the calculation is based on deviations from means, the subtraction of 10 has no effect whatsoever on the finally estimated values of variance and standard deviation. Multiplication by 10, however, means that the initially calculated values for variance and standard deviation are respectively 100 and 10 times too big. These may be corrected at the end of the calculation if needed.

The modified observations of Example 1 are set out in the central part of Table 6.1, Example 1e(b).

The sum of squares of the deviations of individual responses (y) from their grand mean for all doses ($\bar{y}$) is calculated. It is conveniently obtained by the identity

$$S(y - \bar{y})^2 = Sy^2 - (Sy)^2/n \tag{6.1}$$

Using nomenclature analogous to that introduced in Section 2.9,

$$S_{yy} = (93^2 + 72^2 + \cdots + 38^2) - (93 + 72 + \cdots + 38)^2/24$$
$$= 108,458.00 - 95,760.66 = 12,697.34$$

The figure 12,697.34 may be regarded for the present purpose as consisting of two components: a major component arising from difference in dose levels and a minor component representing error. It is the latter that we wish to estimate. This, the (residual) variance of y is obtained by removing the component due to dose difference (the regression component) by means of the expression

$$V(y) = \frac{S_{yy} - b^2(S_{xx})}{n - 2} \tag{6.2}$$

in which $n - 2$ is the degrees of freedom given by

(total no. of observations) $- (1) - (1$ d.f. for regression).

From the data of Example 1, the following values were calculated in Section 2.9

$$S_{xx} = 2.7180, \qquad S_{xy} = 18.2105, \qquad b = S_{xy}/S_{xx} = 6.70$$

In this calculation, as observed values of y have been modified, the values of S_{xy} and b that were obtained using the unmodified observations y must now be increased tenfold. Thus, substituting in Eq. (6.2),

$$V(y) = \frac{12,697.34 - (67.00)^2(2.718)}{(24 - 2)} = 22.57$$

The (residual) standard deviation of modified observations is given by

$$s_y = [V(y)]^{1/2} \tag{6.3}$$

as $(22.57)^{1/2} = 4.751$. In terms of observed zone diameters this represents a standard deviation of 0.475 mm.

6.4   Analysis of Variance to Separate Components Attributable to Various Sources

The foregoing calculation procedure is applicable when test solutions are distributed on the plate in a random manner and when there is no regular pattern of dose. In this example, however, two features of the design make it possible to calculate a more appropriate value for $V(y)$. This new value is lower and more realistic because it isolates and removes additional identifiable sources of variation. These two features of the design are:

(1)   Each of the four test solutions was applied to each plate once and once only, so that each plate may be regarded as a self-contained assay. Because the treatment of all plates is the same, variation of responses between the plates can be identified and isolated in the calculation.

(2)   Doses form a geometrical progression and so log doses are evenly spaced, thus permitting the use of orthogonal polynomial coefficients in the calculation. Consequently variation due to curvature of the response line is isolated and removed.

This is demonstrated in Example 1e(b) by using the technique of "analysis of variance." The relationship with the calculation of Example 1e(a) is also shown.

Example 1e(b):   Estimation of Variance and Standard Deviation of Zone Diamter from a Standard Curve for Streptomycin (*Bacillus subtilis*, Agar Diffusion Assay Method)

Zone sizes are shown in their modified form in Table 6.1. Each row of modified observations corresponds to the zones appearing on one plate. The cumulative totals for all but the last row are not essential. Nevertheless it is useful to record them, as they may be useful in rechecking if arithmetical errors are suspected. The total of all

Table 6.1

Modified Data for Evaluation of the Standard Curve for Streptomycin [Example 1e(b)]

Plate totals	Modified responses to doses (IU/ml)				Cumulative totals	
	80	40	20	10	Sums	Sums of squares
250	93	72	53	32	250	17,606
270	99	78	58	35	520	38,140
242	96	66	45	35	762	54,962
230	87	67	47	29	992	70,070
249	93	73	52	31	1241	87,713
275	96	79	62	38	1516	108,458
1516	564	435	317	200		

observations is obtained at the bottom of both columns 1 and 6. A third check is provided by adding together the totals of columns 2–5.

Analysis of variance.

Total deviation squares: S_{yy} is obtained as in Example 1e(a) and is 12,697.23.

Plate deviation squares: A component of variation due to difference between plates is calculated as the sum of the squares of the deviations of plate totals (sum of zone diameters on a plate) from the mean of all plate totals, divided by the number of zones on each plate. The calculation in this manner is shown in Table 6.2. However, it is much more conveniently calculated in the form analogous to that introduced in Section 2.9 for S_{xx}:

$$\frac{\text{sum of (plate totals)}^2}{\text{number of zones per plate}} - \frac{\text{(grand total)}^2}{\text{total number of zones}}$$

Table 6.2

Computation of Component of Variation Due to Differences between Plates [Example 1e(b)][a]

Plate totals	Deviations from mean plate total	Deviations squared
250	−2.667	7.1129
270	+17.333	300.4329
242	−10.667	113.7849
230	−22.667	513.7929
249	−3.667	13.4469
275	+22.333	498.7629

[a] The mean of all plate totals is 252.667. Sum of squares: 1447.3334; component of variation due to plates (squares for plates): 1447.333/4 = 361.833.

Thus,

$$\frac{250^2 + 270^2 + \cdots + 275^2}{4} - \frac{1516^2}{24} = 96{,}122.50 - 95{,}760.66 = 361.84$$

Treatment deviation squares: In a similar manner a component of variation due to differences between the different treatments (doses in this case) is calculated as

$$\frac{\text{sum of (treatment totals)}^2}{\text{number of zones per treatment}} - \frac{\text{(grand total)}^2}{\text{total number of zones}}$$

which is

$$\frac{564^2 + 435^2 + 317^2 + 200^2}{6} - \frac{1516^2}{24} = 107{,}968.33 - 95{,}760.66 = 12{,}207.67$$

Isolation of regression squares: Using orthogonal polynomial coefficients, this variation due to treatments can be further broken down into components separating the required regression (slope) component from the undesirable components for quadratic and cubic curvature. The coefficients are obtained from mathematical tables such as those of Fisher and Yates (1963).

This part of the calculation is shown in Table 6.3. The figures in the column headed e_i are the sums of the squares of the coefficients in the same row. The figures in the column headed T_i are the sums of the products of the treatment totals in the bottom row and the corresponding coefficients in the column above.

Table 6.3

Breakdown of Variation Due to Treatments into Its Components [Example 1e(b)]

	Orthogonal polynomial coefficients				e_i	T_i	$T_i^2/6e_i$
Regression	+3	+1	−1	−3	20	+1210	12,200.83
Quadratic curvature	+1	−1	−1	+1	4	+12	6.00
Cubic curvature	+1	−3	+3	−1	20	+10	0.83
Treatment total	564	435	317	200			

Thus, for row 1,

$$e_i = +3^2 + 1^2 + (-1)^2 + (-3)^2 = 20$$
$$T_i = (3 \times 564) + (435) + (-317) + (-3 \times 200) = +1210$$

In the final column heading, 6 represents the replication, the number of zones for each treatment. The total of the figures in the last column should equal that previously

Table 6.4

Summary of Analysis of Variance [Example 1e(b)]

Source of variation	Degrees of freedom	Squares	Mean squares
Regression	1	12,200.83	12,200.83
Quadratic curvature	1	6.00	6.00
Cubic curvature	1	0.83	0.83
Subtotal	3	12,207.66	
Treatments	3	12,207.67	
Plates	5	361.84	72.37
Error (by difference)	15	127.83	8.52[a]
Total	23	12,697.34	

[a] From error mean squares,

$$s^2 = 8.52 \quad \text{and} \quad s = (8.52)^{1/2} = 2.92$$

In terms of unmodified observations this represents 0.29 mm.

obtained for "treatment squares," thus providing another check on the calculation. Unless agreement is very close, e.g., a difference only in the second decimal place, then a calculation error is indicated.

The information so far obtained is summarized in Table 6.4 and a figure representing residual (random) error is obtained as the difference between total "squares" and "squares" for identifiable sources. Thus, residual error squares is given by

$$\text{(total squares)} - \text{(treatment squares + plate squares)}$$

as

$$12,697.34 - (12,207.67 + 361.84) = 127.83$$

In the last column of Table 6.4 the figures are obtained by dividing the corresponding squares by the degrees of freedom (d.f.), which are obtained as follows:

$$\text{total} = \text{number of zones less one} = 24 - 1 = 23$$
$$\text{treatments} = \text{number of treatments less one} = 4 - 1 = 3$$
$$\text{plates} = \text{number of plates less one} = 6 - 1 = 5$$
$$\text{error} = \text{(total d.f.)} - \text{(treatments d.f.)} - \text{(plates d.f.)}$$
$$= 23 - 3 - 5 = 15$$

Regression, quadratic curvature, and cubic curvature are all components of variation due to treatments and have one degree of freedom each, thus equaling the number for treatments.

The new figure calculated for variance of y is the mean square for error obtained as $127.83/15 = 8.52$. Standard deviation of modified observations is obtained by Eq. (6.3) as

$$s_y = (8.52)^{1/2} = 2.92$$

In terms of observed zone diameters this represents a standard deviation of 0.292 mm.

The relationship between calculations 1e(a) and 1e(b) may be seen by calculating from the figures given in Table 6.4:

$$\frac{\text{(total squares)} - \text{(regression squares)}}{\text{(total d.f.)} - \text{(regression d.f.)}}$$

$$= \frac{12{,}697.34 - 12{,}200.83}{23 - 1} = \frac{496.51}{22} = 22.57$$

which is identical with the value obtained in Example 1e(a).

Apart from separating the error term from which confidence limits can be calculated, analysis of variance serves another purpose. It permits a comparison of the extent of variation attributable to various sources. On inspection of the mean squares column of Table 6.4, it is seen that the values corresponding to both forms of curvature are small compared with that for "error." This suggests that curvature is not significant.

However, the figure for mean squares for differences between plates is large compared with the mean squares for residual error. This might suggest even to the nonstatistician that differences between plates are a real source of variation not attributable to residual error. This comparison can be put on a quantitative basis by use of variance ratio in the "F test."

The ratio of the larger to the smaller of the two variances (mean squares) is calculated. The significance of this ratio is then compared with values given in the variance ratio tables of Fisher and Yates (1963) or the abridged version given in Appendix 10. These tables define levels of significance for certain ratios, taking into consideration the degrees of freedom associated with each variance.

This will be made clearer by an example.

In Example 1e(b) we have found:

> plate mean squares $= 72.37$ with 5 d.f.
>
> error mean squares $= 8.52$ with 15 d.f.

Their variance ratio is $72.37/8.52 = 8.49$. Reference is made to the variance ratio tables as follows:

n_1 is the number of degrees of freedom of the source of variation having the greater mean square.

n_2 is the number of degrees of freedom of the source of variation having the lesser mean square.

Thus in this case, $n_1 = 5$ and $n_2 = 15$.

First referring to the 5% table, the value corresponding to column $n_1 = 5$ and row $n_2 = 15$ is 2.90.

Corresponding values in the 1 and 0.1% tables are 4.56 and 7.57, respectively.

As the observed ratio 8.49 is even greater than 7.57, the evidence strongly suggests that the differences between plates cannot be attributed to random variation.

Thus the intuitive reasoning of the nonstatistician is in this case shown to be correct. In many cases however, mere inspection of the values of mean squares does not give any clear indication of significance of sources of variation. It is in such cases that the variance ratio test is of real value.

It is appropriate at this juncture to elaborate on the meaning of the statistical term "significant." The usual convention is to regard a probability of 95% as "significant" and a probability of 99% or more as "highly significant." These terms refer to the probability of a postulated effect being real. They are not synonymous with "importance" in the context of the purpose of the assay. Thus an apparent curvature might be classified as nonsignificant if the random error of replicate observations were so great as to cast doubt on the reality of the curvature. In contrast, the same apparent degree of curvature might be classified as highly significant if random error were very small.

It is shown in Chapter 8 that curvature may often be of little importance. However, a statistically significant departure from parallelism would be cause for concern and investigation.

Returning to consideration of Example 1e(b), differences between plates have been shown to be highly significant. However, in this experimental design, each plate can be considered as a self-contained unit including one dose of each treatment. If it could be assumed that differences between overall plate responses affected all treatments equally, then this design would correct perfectly for plate variation. Consideration of the theory of zone formation, Section 2.2, reveals that any factor influencing zone size also influences slope and so perfect correction will not be attained. In practice the correction is very good as may be seen in Example 1e by a consideration of the substantial term for plate squares in Table 6.4, which leads to a corresponding reduction in the term for error squares.

In the normal case, separation of the term for plate squares also leads to a reduction in the figure for *mean* error squares (even though degrees of freedom are reduced).

In this particular example, the error mean squares term s^2 has been used only to calculate standard deviation of the zone size. This in itself is not of direct use. However, when the object of the practical exercise is to estimate the potency of an unknown sample, s^2 is of fundamental importance in the assessment of confidence limits to the potency estimate.

6.5 Evaluation of a Simple Two Dose Level Assay

The principles of analysis of variance are now applied to a simple example of a comparison of a sample with a standard at two dose levels, that is, to Example 2. As in the previous case, observations have been modified, this time multiplying by 10 then subtracting 150.

Thus, 20.1 becomes $(10 \times 20.1) - (150) = 51$.

Example 2e: The Agar Diffusion Assay of Streptomycin by Simple 2 + 2 Design (Modified responses are presented in Table 6.5)

Total deviation squares (from Eq. (6.1)):

$$S_{yy} = (51^2 + 59^2 + \cdots + 8^2) - 805^2/24 = 38,129.00 - 27,001.04 = 11,127.96$$

Table 6.5

Data of Example 2e—Modified Responses to High and Low Doses[a]

	Sample		Standard		
Dose	High	Low	High	Low	Plate totals
Plate number					
1	51	7	48	3	109
2	59	15	57	9	140
3	59	14	54	16	143
4	58	17	60	13	148
5	56	18	52	14	140
6	49	15	53	8	125
Treatment total	332	86	324	63	805

[a] Test solutions had nominal potencies of 80 and 20 IU/ml, respectively.

Treatment deviation squares (often written just as "treatment squares") is calculated exactly as in the previous example but does not have exactly the same meaning. In Example 1e, each treatment was a different dose of the same preparation. In this example, two treatments correspond to two doses of standard and the other treatments to two doses of sample. Thus the breakdown of the treatment squares that follows later in the calculation differs from that of Example 1e.

Treatment squares:

$$\frac{332^2 + 86^2 + 324^2 + 63^2}{6} - \frac{805^2}{24} = 37{,}760.83 - 27{,}001.04 = 10{,}759.79$$

Preparation squares:

$$\frac{S\,(\text{preparation totals})^2}{\text{number of zones per preparation}} - \frac{(Sy)^2}{n}$$

$$= \frac{(332 + 86)^2 + (324 + 63)^2}{12} - \frac{805^2}{24} = 27{,}041.08 - 27{,}001.04 = 40.04$$

Slope or regression squares may be obtained in this particular case by

$$\frac{S\,(\text{dose totals})^2}{\text{number of zones per dose level}} - \frac{(Sy)^2}{n}$$

$$= \frac{(332 + 324)^2 + (86 + 63)^2}{12} - \frac{805^2}{24} = 37{,}711.42 - 27{,}001.04 = 10{,}710.38$$

This procedure is only applicable to two dose level assays in which it is impossible to detect any deviations from linearity. If the same procedure is applied to multidose level assays, a gross figure is obtained that includes deviations from linearity. This would be analogous to the treatment squares of Example 1e(b), in which the gross figure is seen to comprise regression, quadratic curvature, and cubic curvature components.

The value for regression squares may be obtained more conveniently in the case of two dose level assays (and three dose level assays in which log dose intervals are equal) as:

$$[(\text{high response totals}) - (\text{low response totals})]^2/n$$

$$= [(332 + 324) - (86 + 63)]^2/24 = (507)^2/24 = 10{,}710.38$$

The slopes for responses of sample and standard may differ. Parallelism (deviations) squares is a measure of the deviation of slopes from one another and is obtained as

$$\left[\frac{S\,(\text{preparation slopes})^2}{\text{number of zones per preparation}}\right] - \left[\text{regression squares}\right]$$

$$= \frac{(332 - 86)^2 + (324 - 63)^2}{12} - 10{,}710.38 = 10{,}719.75 - 10{,}710.38 = 9.37$$

The same values of squares for preparations, regression, and parallelism may also be obtained by expressing the calculation in terms of orthogonal polynomial coefficients as shown in Table 6.6.

Table 6.6

Part of the Analysis of Variance of Example 2e

Sources	Orthogonal polynomial coefficients				e_i	T_i	$T_i^2/6e_i$
Preparations	+1	+1	−1	−1	4	+31	40.04
Regression	+1	−1	+1	−1	4	+507	10,710.37
Parallelism	+1	−1	−1	+1	4	−15	9.37
Totals	332	86	324	63			

The remaining identifiable source of variation is the plates. *Plate squares* is calculated as in Example 1e(b) as

$$\frac{S\,(\text{plate totals})^2}{\text{number of zones per plate}} - \frac{(Sy)^2}{n}$$

$$= \frac{109^2 + 140^2 + 143^2 + 148^2 + 140^2 + 125^2}{4} - 27{,}001.04$$

$$= 27{,}264.75 - 27{,}001.04 = 263.7$$

The values of all sums of squares are summarized in Table 6.7.

Table 6.7

Summary of the Analysis of Variance (Example 2e)

Source of variation	Degrees of freedom	Sum of squares	Mean squares
Preparations	1	40.04	
Regression	1	10,710.38	
Parallelism (deviations)	1	9.37	9.37
Subtotal	3	10,759.79	
Treatments	3	10,759.79	
Plates	5	263,71	52.74
Residual error	15	104.46	6.96[a]
Total (all sources)	23	11,127.96	

[a] From error mean squares, $s^2 = 6.96$; $s = (6.96)^{1/2} = 2.64$. In terms of unmodified observations, this represents 0.26 mm.

As the basic assumption of these calculations is that standard and sample response lines are parallel, it is important to check that they are in fact parallel. Expressing it rather more realistically, it is important that deviations from parallelism should be no greater than can be attributed to random error. This can be checked by using variance ratios that were introduced in Example 1e(b).

In this case, to determine whether the deviation from parallelism is acceptable, the variance ratio is calculated:

$$\frac{\text{parallelism (deviations) mean squares}}{\text{residual error mean squares}} = \frac{9.37}{6.96} = 1.35.$$

Reference to the 5% table for variance ratio (Appendix 10) shows the limiting ratio for $n_1 = 1$ and $n_2 = 15$ to be 4.5. As the ratio found (1.35) is less than this limiting ratio, the evidence from this assay does not strongly contradict the assumption of parallelism, and hence by convention this is accepted as evidence sufficient to retain it.

The corresponding limiting value from the 20% table for variance ratio is 1.8. As this too is higher than the variance ratio found, the evidence for retention of the assumption of parallelism is strengthened.

It is self-evident that if the standard error s of a single response measurement were not appreciably smaller than the slope b of the response line, then the assay would be of little value. This idea is expressed quantitatively by g, the index of significance of the slope, which is defined as

$$g = \frac{s^2 t^2}{b^2 S_{xx}} \tag{6.4}$$

where s^2 is the residual error mean squares, b the regression coefficient determined in the calculation of potency, S_{xx} the sum of squares of deviation of all log doses from their grand mean, and t the "Student's t," which may be described simply as a factor relating estimate of standard deviation and degrees of freedom to probability. It is obtained from tables such as those of Fisher and Yates (1963) or from Appendix 9.

Values of g less 0.1 indicate a highly significant slope—an essential condition for a good assay—whereas at values approaching 1.0, slope is too poor for a meaningful assay. For values of g exceeding 0.1, Finney (1964) quotes the general formula for the confidence limits of $M - \bar{x}_S + \bar{x}_T$ as

$$\left[M - \bar{x}_S + \bar{x}_T \pm \frac{ts}{b} \left\{ (1 - g) \left(\frac{1}{N_S} + \frac{1}{N_T} + \frac{(M - \bar{x}_S + \bar{x}_T)^2}{S_{xx}} \right) \right\}^{1/2} \right] \div (1 - g) \tag{6.5a}$$

where M is the logarithm of estimate of potency ratio, unknown: standard, N_S the number of standard observations, N_T the number of sample observations, $\bar{x}_S$ the mean standard log dose, and $\bar{x}_T$ the mean sample log dose.

However, when g is less than 0.1 an approximate formula may be used in place of Eq. (6.5a). It is obtained by substituting in (6.5a) $g = 0$. Thus confidence limits may be obtained from

$$M - \bar{x}_S + \bar{x}_T \pm \frac{ts}{b} \left\{ \frac{1}{N_S} + \frac{1}{N_T} + \frac{(M - \bar{x}_S + \bar{x}_T)^2}{S_{xx}} \right\}^{1/2} \tag{6.5b}$$

Expressions (6.5a) and (6.5b) are analagous to the two forms quoted in the International, European, and United States Pharmacopeias, which use different nomenclatures and give confidence limits as a percentage of the estimated potency.

Experience indicates that in *microbiological* assays the value of g is always very much less than 0.1 and so the form (6.5b) is applicable. The procedure adopted in this work is to obtain confidence limits via the intermediate steps of variance of M and standard error of M.

The variance $V(M)$ of the logarithm of the estimated potency M is given by the following expression, which is an approximation very simply related to (6.5b):

$$V(M) = \frac{s^2}{b^2}\left[\frac{1}{N_S} + \frac{1}{N_T} + \frac{(M - \bar{x}_S + \bar{x}_T)^2}{S_{xx}}\right] \tag{6.6}$$

Then s_M, the standard error of M, is obtained by

$$s_M = [V(M)]^{1/2} \tag{6.7}$$

Logarithms of percentage confidence limits are given by

$$2 \pm ts_M \tag{6.8}$$

These expressions are now applied to the data of Example 2e.

To calculate $V(M)$, the following values are needed:

$s^2 = 6.96$ the residual error mean squares from the analysis of variance

$t = 2.13$ this value corresponds to 15 degrees of freedom at $P = 0.95$ and is obtained from tables (see Appendix 9)

$b = 70.17$ from the potency calculation, Example 2, this is 7.017 mm, but it must be multiplied by 10 in conformity with the modified zone sizes

$S_{xx} = 2.1744$ this derivation is given below

The sum of the squares of deviations of individual log doses from the mean log dose may be obtained by Eq. (2.7). However, in this particular example, as the nominal dose levels for sample and standard are equal, the dose intervals on the logarithmic scale are equal, and the number of doses at each level are equal, S_{xx} may be calculated more simply in its original form as $S(x - \bar{x})^2$.

The interval between the two dose levels is 0.602 (log scale), and all deviations are evenly distributed at either $+0.301$ or -0.301 from the mean.

Thus, the required figure is most easily obtained as

$$S_{xx} = 12[(+0.301)^2 + (-0.301)^2] = 2.1744$$

Now substituting these values in Eq. (6.4),

$$g = \frac{6.96 \times (2.13)^2}{(70.17)^2 \times 2.1744} = 0.0029$$

As g is less than 0.1, $V(M)$ is calculated using the simplified formula, Eq. (6.6). In this example $\bar{x}_S = \bar{x}_T$ and so variance of M is obtained as

$$V(M) = \frac{6.96}{(70.17)^2}\left[\frac{1}{12} + \frac{1}{12} + \frac{(0.0368)^2}{2.1744}\right]$$

$$= 0.0014135[0.166,667 + 0.000,623] = 0.0002365$$

The standard error of M is given by Eq. (6.7) as

$$s_M = [0.0002365]^{1/2} = 0.01538$$

Logarithm of percentage confidence limits are then obtained by Eq. (6.8). In this case log percent confidence limits are

$$2 \pm 2.13 \times 0.01538 = 1.9672 \quad \text{to} \quad 2.0328.$$

Corresponding confidence limits are 92.7–107.8% of the estimated potency. Sample potency was estimated in Example 2 (Chapter 2) as 741 IU/mg and so the sample may now be reported as

$$741 \quad \text{IU/mg, limits 687 to 800} \quad \text{IU/mg} \quad (P = 0.95)$$

6.6 An Alternative Method for Obtaining Confidence Limits

Confidence limits have been expressed in a somewhat different form by Lees and Tootill (1955b). For every assay design a separate simple formula is derived. For example, for the 2 + 2 assay design described for streptomycin (Example 2)

$$\text{Standard error (as \%)} = 100 \times \ln 4 \left[\frac{\text{error mean squares}}{\text{regression squares}}\right]^{1/2} \quad (6.9)$$

Substituting the values from Example 2,

$$\text{Standard error (\%)} = 100 \times 1.3863 \left[\frac{6.96}{10,710.38}\right]^{1/2} = 3.53\%$$

Thus for $P = 0.95$, confidence limits are about $100 \pm 2 \times 3.53\%$ or 92.9–107.1%. For most practical purposes this result is close enough to the value first calculated.

Similar formulas derived by this method for other designs are given by Lees and Tootill (1955).

6.7 Evaluation of a Multiple Two Dose Level Assay

The inclusion of an additional sample in the 2 + 2 assay of streptomycin by the petri dish method requires only minor modifications to the statistical treatment. This is shown in Example 3e.

Example 3e: Agar Diffusion Assay of Streptomycin by 2 + 2 Design for Two Samples

Zone sizes modified as in Example 2e are given in Table 6.8. Then following the same procedures as before:

Total deviation squares (from Eq. (6.1)):

$$S_{yy} = 55{,}801.00 - (1179)^2/36 = 55{,}801.00 - 38{,}617.25 = 17{,}188.75$$

Table 6.8

Data of Example 3e; Modified Responses to the Two Dose Levels High and Low[a]

	Sample 1		Sample 2		Standard		Plate totals
Dose:	High	Low	High	Low	High	Low	
Plate number							
1	51	7	53	9	48	3	171
2	59	15	55	8	57	9	203
3	59	14	55	9	54	16	207
4	58	17	52	12	60	13	212
5	56	18	55	8	52	14	203
6	49	15	51	7	53	8	183
Treatment totals	332	86	321	53	324	63	1179

[a] Test solutions had nominal potencies of 80 and 20 IU/ml, respectively.

Treatment squares:

$$\frac{332^2 + 86^2 + 321^2 + 53^2 + 324^2 + 65^2}{6} - \frac{1179^2}{36}$$

$$= 55{,}402.50 - 38{,}612.25 = 16{,}790.25$$

Preparation squares:

$$\frac{(332 + 86)^2 + (321 + 53)^2 + (324 + 63)^2}{12} - \frac{1179^2}{36}$$

$$= 38{,}697.42 - 38{,}612.25 = 85.17$$

Regression squares:

$$[(332 + 321 + 324) - (86 + 53 + 63)]^2/36$$
$$= (775)^2/36 = 16{,}684.03$$

Parallelism (deviations) squares:

$$\frac{(332 - 86)^2 + (321 - 53)^2 + (324 - 63)^2}{12} - \frac{775^2}{36}$$

$$= 16{,}705.08 - 16{,}684.03 = 21.05$$

Plate squares:

$$\frac{171^2 + 203^2 + 207^2 + 212^2 + 203^2 + 183^2}{6} - \frac{1179^2}{36}$$

$$= 38{,}823.50 - 38{,}612.25 = 211.25$$

Sums of squares are summarized in Table 6.9.

Table 6.9

Summary of the Analysis of Variance (Example 3e)

Source of variance	Degrees of freedom	Sums of squares	Mean squares
Preparation	2	85.17	
Regression	1	16,684.03	
Parallelism (deviations)	2	21.05	10.52
Subtotal	5	16,790.25	
Treatments	5	16,790.25	
Plates	5	211.25	
Residual variation (error)	25	187.25	7.49[a]
Total (all sources)	35	17,188.75	

[a] From error mean squares

$$s^2 = 7.49 \quad \text{and} \quad s = (7.49)^{1/2} = 2.74$$

In terms of unmodified observations this represents 0.27 mm.

The variance ratio for parallelism/residual error is $10.52/7.49 = 1.41$. Reference to the tables shows that for $n_1 = 2$ and $n_2 = 25$, the limiting value is 3.38 (5% table). There is therefore no significant deviation from parallelism.

To determine the confidence limits, g is first evaluated thus:

$s^2 = 7.49$ residual error mean squares from analysis of variance

$t = 2.06$ obtained from table for $P = 0.95$ and 25 degrees of freedom for error mean squares

$b = 71.51$ from the potency calculation (Example 3), $b = 7.151$; this value is multiplied by 10 in conformity with the modification of the zone sizes

$S_{xx} = 3.2616$ obtained as $18[(0.301)^2 + (-0.301)^2]$

From Eq. (6.4),

$$g = \frac{7.49 \times (2.06)^2}{(71.51)^2 \times 3.2616} = 0.0019$$

As g is less than 0.1, $V(M)$ is calculated for Sample 1 by Eq. (6.6) as:

$$V(M) = \frac{7.49}{(71.51)^2}\left[\frac{1}{12} + \frac{1}{12} + \frac{(0.036)^2}{3.2616}\right]$$
$$= 0.0014647[0.166,667 + 0.000,400]$$
$$= 0.000,244,7$$

The standard error of M is then obtained from Eq. (6.7):

$$s_M = [0.000,244,7]^{1/2} = 0.01564$$

Logarithms of percent confidence limits are given by Eq. (6.8):

$$2 \pm 2.06 \times 0.01564 = 1.9678 \quad \text{to} \quad 2.0322.$$

Percent confidence limits of the estimated potency are obtained by taking antilogs and are 92.9–107.7%. The potency of Sample 1 in Example 2 of Chapter 2 was estimated as 740 IU/mg. It may be reported therefore as

$$740 \quad \text{IU/mg, limits 687 to 797} \quad \text{IU/mg} \qquad (P = 0.95)$$

In a similar manner, confidence limits were calculated for Sample 2, which had an estimated potency of 9.65 IU/ml. This was reported as

$$9.65 \quad \text{IU/ml, limits 8.95 to 10.4} \quad \text{IU/ml} \qquad (P = 0.95)$$

The reader may calculate the latter himself, and in doing so verify the negligible effect of M on the value of $V(M)$ provided that M is small.

6.8   Evaluation of a Three Dose Level Assay

The statistical evaluation of a three dose level balanced assay is shown using Example 4e, the assay of penicillin using a set of seven petri dishes.

In this case, two more sources of possible variation appear: (i) curvature of the combined log dose–response line for standard and sample, and (ii) differences in the extent of curvature between the two preparations. These are isolated by the use of orthogonal polynomial coefficients.

Example 4e: Assay of Penicillin by 3 + 3 Design Using Small Plates

Zone diameters are modified for convenience of calculation by first multiplying by 10 and then subtracting 100. The modified values are shown in Table 6.10.

Total deviation squares (from Eq. (6.1)):

$$S_{yy} = 380,796 - (3764)^2/42 = 380,796 - 337,326 = 43,470$$

Table 6.10

Data of Example 4c, Modified Responses to the Three Dose Levels, High, Medium, and Low[a]

	Sample			Standard			
Dose:	High	Medium	Low	High	Medium	Low	Plate totals
	144	107	74	140	104	70	639
	122	93	49	127	97	49	537
	123	80	50	120	86	50	509
	122	90	48	124	83	46	513
	126	78	44	123	80	47	498
	130	93	45	133	91	44	536
	124	94	50	125	90	49	532
Treatment totals	891	635	360	892	631	355	3764

[a] Test solutions had nominal potencies of 1.00, 0.50, and 0.25 IU/ml, respectively.

Treatment squares:

$$\frac{891^2 + 635^2 + 360^2 + 892^2 + 631^2 + 355^2}{7} - \frac{(3764)^2}{42}$$

$$= 378{,}079.4 - 337{,}326.0 = 40{,}753.4$$

Plate squares:

$$\frac{639^2 + 537^2 + 509^2 + 513^2 + 498^2 + 536^2 + 532^2}{6} - \frac{(3764)^2}{42}$$

$$= 339{,}544.0 - 337{,}326.0 = 2218.0$$

Squares corresponding to preparations, regression, deviations from parallelism between preparations, curvature, and deviations between preparations as regards curvature (opposed curvature) are all obtained by the use of polynomial coefficients as illustrated in Table 6.11. Squares corresponding to all sources of variation are summarized in Table 6.12.

It will be seen in Table 6.11 that the figures in the column headed $T_i^2/7e_i$ for the first three lines are identical with those that could have been obtained alternatively by the methods shown previously in Examples 2e and 3e.

Thus for line *b*, the squares for regression may be calculated as

$$[S(H) - S(L)]^2/28 = [1783 - 715]^2/28 = 40{,}736.57$$

Inspection shows that mean squares for parallelism, curvature, and opposed curvature are all smaller than that for residual error, and so without recourse to the variance ratio tables, all three are clearly not significant. It will also be observed that the mean squares for plates is highly significant. As explained for Example 1e(b), this does not

Table 6.11

Part of the Analysis of Variance of Example 4e

Doses:	Sample			Standard			e_i	T_i	$7e_i$	$T_i^2/7e_i$
	High	Medium	Low	High	Medium	Low				
Preparations	+1	+1	+1	−1	−1	−1	6	8	42	1.524
Regression	+1	0	−1	+1	0	−1	4	1068	28	40,736.570
Parallelism	+1	0	−1	−1	0	+1	4	−6	28	1.286
Curvature	+1	−2	+1	+1	−2	+1	12	−34	84	13.762
Opposed curvature	+1	−2	+1	−1	+2	−1	12	−4	84	0.190
Treatment totals	891	635	360	892	631	355				40,753.332

Table 6.12

Summary of the Analysis of Variance (Example 4e)

Source of variation	Degrees of freedom	Squares	Mean squares
Preparations	1	1.542	
Regression	1	40,236.570	
Parallelism (deviations)	1	1.286	1.286
Curvature	1	13.762	13.762
Opposed curvature	1	0.190	0.190
Subtotal	5	40,753.332	
Treatments	5	40.753.4	
Plates	6	2,218.0	369.66
Error (by difference)	30	498.6	16.62[a]
Total (all sources)	41	43,470.0	

[a] From error mean squares

$$s^2 = 16.62 \quad \text{and} \quad s = (16.62)^{1/2} = 4.08$$

In terms of unmodified observations this represents 0.41 mm.

in any way invalidate the assay since the design very largely corrects for plate differences. However, it may be inferred that conditions of test are not well standardized. This perhaps acts as a general indicator that practical operations should be better regulated.

As there is so far no cause to doubt the validity of this assay, the calculation may be continued by evaluating g:

$$S_{xx} = 14[(0.301)^2 + (-0.301)^2] = 2.5368$$
$$b = 12.672 \times 10 = 126.72 \quad \text{(from Example 4)}$$
$$t = 2.04 \quad \text{(for} \quad P = 0.95 \text{ and 30 d.f.)}$$

by Eq. (6.4),

$$g = \frac{16.62 \times (2.04)^2}{(126.72)^2 \times 2.5368} = 0.0017$$

As g is small, calculate the variance of M using the approximate Eq. (6.6):

$$V(M) = \frac{16.62}{(126.72)^2} \left[\frac{1}{21} + \frac{1}{21} + \frac{(0.0031)^2}{2.5368} \right] = 0.00009858$$

The standard error of M is obtained from Eq. (6.7) as

$$s_M = [0.00009858]^{1/2} = 0.009929$$

Logarithms of percent confidence limits are given by Eq. (6.8) as

$$2 \pm 2.04 \times 0.009929 = 1.9797 \quad \text{to} \quad 2.0203$$

Confidence limits are given by the antilogs as 95.4–104.8% for $P = 0.95$.

As the potency was determined as 1607 IU/mg, these limits are 95.4 and 104.8% of 1607, respectively. Thus the potency may be reported as 1607 IU/mg with limits of 1533 and 1684 IU/mg at 95% probability level.

6.9 Evaluation of a Large Plate Assay (Latin Square Design)

The analysis of variance for two dose level assays using Latin square designs differs slightly from that of the two dose level assay using petri dishes (Examples 2e and 3e).

In Chapter 2 the special characteristics of large plates were discussed. It was seen that zone size could be influenced by time of application of the solution to the plate, edge effects, temperature gradients, possible failure to level the plate exactly, and uneven dispersion of the test organism.

A combination of some or all of these influences is revealed as differences between rows and between columns, despite each row and column containing each test solution once only.

Whereas in the petri dish assay a component of variation attributable to "plates" was isolated, in the case of the Latin square design two components for "rows" and "columns" must be isolated. To achieve this, responses must be set out in the same pattern as that in which they occurred on the plate.

The procedure is illustrated using Example 5, the assay of neomycin.

Example 5e: Assay of Neomycin by 2 + 2 Assay for Three Samples Compared with One Standard Using an 8 × 8 Latin Square Design

To reduce the labor in computation, observed zone diameters were modified by multiplying by 10 and then subtracting 190. These modified responses are shown in Table 6.13 in columns according to treatments (test solution numbers). They are then

Table 6.13

Data of Example 5e[a]

	Sample 1		Sample 2		Sample 3		Standard	
Dose:	1	2	3	4	5	6	7	8
	43	21	37	6	43	20	41	15
	44	12	39	11	39	15	39	11
	39	11	36	5	39	19	40	13
	41	16	35	6	42	17	37	15
	46	13	38	13	46	19	40	14
	44	21	41	16	50	24	46	19
	49	23	48	15	50	14	47	19
	51	24	44	17	55	30	50	24
Treatment total	357	141	318	89	364	158	340	130

[a] Modified responses tabulated according to treatment number. Odd numbers refer to high doses and even numbers to low doses. Test solutions were of nominal potencies 60 and 15 U/ml, respectively.

Table 6.14

Data of Example 5e[a]

			Modified responses					Row total
43	20	15	41	37	6	43	21	226
39	44	39	15	12	11	39	11	210
19	13	36	39	5	39	11	40	202
15	35	16	6	41	37	17	42	209
46	13	46	13	40	19	38	14	229
46	50	24	41	19	21	16	44	261
15	23	49	19	50	48	47	14	265
24	50	17	51	30	55	24	44	295
Column total	247	248	242	225	234	236	235	230	1897

[a] Modified responses set out in the same arrangement as that in which they appeared on the plate according to the Latin square design.

also set out in Table 6.14 in the same arrangement as that in which they occurred in the Latin square design, Fig. 2.14.

Analysis of Variance:

Total deviation squares (from Eq. (6.1)):

$$S_{yy} = 69{,}585 - (1897)^2/64 = 69{,}585 - 56{,}228.26 = 13{,}356.74$$

From the tabulated treatment totals, sums of squares are obtained for treatments, preparations, regression, and parallelism as follows:

Treatment squares:

$$\frac{357^2 + 141^2 + 318^2 + 89^2 + 364^2 + 158^2 + 340^2 + 130^2}{8} - \frac{(1897)^2}{64}$$

$$= 68{,}291.87 - 56{,}228.26 = 12{,}063.61$$

Preparation squares:

$$\frac{(357 + 141)^2 + (318 + 89)^2 + (364 + 158)^2 + (340 + 130)^2}{16} - \frac{(1897)^2}{64}$$

$$= 56{,}689.81 - 56{,}228.26 = 461.55$$

Regression squares:

$$\frac{[(357 + 318 + 364 + 340) - (141 + 89 + 158 + 130)]^2}{64}$$

$$= \frac{741{,}321}{64} = 11{,}583.14$$

Parallelism (deviations) squares:

$$\frac{(357 - 141)^2 + (318 - 89)^2 + (364 - 158)^2 + (340 - 130)^2}{16} - 11{,}583.14$$

$$= 11{,}602.06 - 11{,}583.14 = 18.92$$

From the values tabulated for totals of row responses and column responses of the Latin square design, sums of squares for rows and columns are obtained as follows:

Row squares:

$$\frac{226^2 + 210^2 + 202^2 + 209^2 + 229^2 + 261^2 + 265^2 + 295^2}{8} - \frac{(1897)^2}{64}$$

$$= 57{,}184.12 - 56{,}228.26 = 955.86$$

Column squares:

$$\frac{247^2 + 248^2 + 242^2 + 225^2 + 234^2 + 236^2 + 235^2 + 230^2}{8} - \frac{(1897)^2}{64}$$

$$= 56{,}284.87 - 56{,}228.26 = 56.61$$

The analysis of variance is summarized in Table 6.15.

The value of squares for parallelism is smaller than squares for error. Without recourse to the variance ratio tables it is clear that apparent deviations from parallelism may be neglected. Thus far there is no cause to doubt the validity of the assay.

Table 6.15

Summary of Analysis of Variance (Example 5e)

Sources of variation	Degrees of freedom	Squares	Mean squares
Preparations	3	461.55	153.85
Regression (slope)	1	11,583.14	
Parallelism (deviations)	3	18.92	6.31
Subtotal	7	12,063.61	
Treatments	7	12,063.61	
Rows	7	955.86	136.55
Columns	7	56.61	8.09
Residual error	42	280.66	6.68[a]
Total (all sources)	63	13,356.74	

[a] From error mean squares

$$s^2 = 6.68 \quad \text{and} \quad s = (6.68)^{1/2} = 2.59$$

In terms of unmodified observations this represents 0.26 mm.

It will be noted that the value for row mean squares is much greater than that for column mean squares. This is a reflection of the differences in mean diffusion times that were described in Section 2.10.

The ratio

$$\frac{\text{row mean squares}}{\text{error mean squares}} = \frac{136.6}{6.7} = 20.5$$

Comparing this value with the appropriate limiting values in the variance ratio tables, putting $n_1 = 7$ and $n_2 = 42$ suggests a highly significant difference between rows. However, even though the difference between rows may be significant, distribution of doses in accordance with the Latin square design coupled with the correct plating technique ensures that bias to estimated potency is minimal.

Proceeding to the determination of confidence limits of the potency estimate, calculate g by Eq. (6.4):

$$S_{xx} = 32[(0.301)^2 + (-0.301)^2] = 5.7985$$
$$b = 4.4685 \times 10 = 44.685 \quad \text{(from Example 5)}$$
$$t = 2.02 \quad \text{(for } P = 0.95 \text{ and 42 d.f.)}$$

Thus

$$g = \frac{6.68 \times (2.02)^2}{(44.685)^2 \times 5.7985} = 0.0024$$

As g is small, calculate $V(M)$ by the approximate Eq. (6.6). For Sample 1,

$$V(M_1) = \frac{6.68}{(44.685)^2}\left[\frac{1}{16} + \frac{1}{16} + \frac{(0.0392)^2}{5.7985}\right] = 0.000419$$

By Eq. (6.7),

$$s_{M_1} = [0.000419]^{1/2} = 0.02047$$

Log percent confidence limits are given by Eq. (6.8) as

$$2 \pm 2.02 \times 0.02047 = 1.9581 \quad \text{to} \quad 2.0414$$

The corresponding percent confidence limits are therefore 90.9–110.0% ($P = 0.95$). In Chapter 2, Example 5, potency was estimated as 607 U/mg, and so it may be reported as 607 U/mg, limits 551–668 U/mg ($P = 0.95$). Similarly, confidence limits may be calculated for Samples 2 and 3. The different values of M_1, M_2, and M_3 when substituted in Eq. (6.6) have a quite insignificant effect, and percent confidence limits are in all three cases 90.9–110.0%. Thus, Sample 2 was reported as 458 U/mg, limits 416–504 U/mg ($P = 0.95$) and Sample 3 as 723 U/mg, limits 657–795 U/mg ($P = 0.95$).

The microbiological assay of neomycin is well known for the poor reproducibility of results. A major factor contributing to this, which is illustrated in the preceding example, is the low value of the regression coefficient. In this case $b = 4.469$ mm, which leads to low precision. However, probably a more serious difficulty is the variable composition of the commercially available antibiotic. This has been discussed in Chapters 1 and 5.

6.10 Evaluation of a Large Plate Assay (Quasi-Latin Square Design)

In the case of assays employing quasi-Latin square designs, there is an additional factor to be taken into consideration. For each preparation the two different doses are allocated a pair of numbers consisting of an odd number and the next higher even number, e.g., 5 and 6, or 7 and 8, but not 6 and 7. By this means each row and each column of the design contains one dose level of each preparation, thus minimizing bias due to positional effects. (This contrasts with the true Latin square design used in Example 5, in which both dose levels of each preparation occurred in each row and each column, giving better compensation for positional effects.)

A consequence of this arrangement is that there are two groups of rows and two groups of columns. These can be identified as those containing test solution 1, group A, and those containing test solution 2, group B. Considering the columns, it can be seen from Fig. 2.14 that group A columns contain high dose test solutions 1, 5, 11, and 15, as well as low dose test solutions 4, 8, 10, and 14. Group B columns contain all the corresponding members of the pairs, i.e., low dose test solutions 2, 6, 12, and 16, as well as high dose test solutions 3, 7, 9, and 13.

If slopes for all eight preparations were identical, then the sum of all A columns minus the sum of all B columns would be zero. When this difference is not zero it might appear to indicate differences in slopes. However, the influence of position on zone size is a known fact and so it can be taken that this discrepancy from zero is a measure of the false contribution to slope differences, which arise from positional differences.

Accordingly, row and column corrections are made to the figure initially calculated for parallelism squares, which then loses two degrees of freedom.

In a similar way, since the row squares and column squares as calculated initially contain a component reflecting their A and B groups, they too must be corrected and each lose one degree of freedom.

The row and column corrections make a positive contribution to the treatment totals with one degree of freedom each.

It may be seen from Fig. 2.15 that the test solution numbers in group A columns are not identical with those of group A rows. This fact, however, is not relevant to the calculation.

The calculation pattern is illustrated in Example 6e, the assay of vitamin B_{12}.

Example 6e: Agar Diffusion Assay of Vitamin B_{12} by Multiple $2 + 2$ Design Using a Large Plate and an 8×8 Quasi-Latin Square Design

For convenience, observed zone diameters have been modified by multiplying by 10 and then subtracting 180. Modified observations are first arranged in Table 6.16 according to test solution number, and then in Table 6.17 as they appeared on the plate in accordance with the quasi-Latin square design.

Analysis of variance

Total deviation squares (from Eq. (6.1)):

$$S_{yy} = 155,385 - (2791)^2/64 = 155,385 - 121,713.76 = 33,671.24$$

From the tabulated treatment totals, sums of squares are obtained for treatments, preparations, and regression:

Treatment squares:

$$\frac{261^2 + 76^2 + 247^2 + \cdots + 265^2 + 96^2}{4} - \frac{2791^2}{64}$$

$$= \frac{620,297}{4} - \frac{2791^2}{64}$$

$$= 155,074.25 - 121,713.76 = 33,360.49$$

Preparation squares:

$$\frac{337^2 + 321^2 + \cdots + 361^2}{8} - \frac{2791^2}{64}$$

$$= \frac{983,349}{8} - \frac{2791^2}{64}$$

$$= 122,918.62 - 121,713.76 = 1,204.86$$

Table 6.16

Assay of Vitamin B_{12}[a]

Samples:	A		B		C		D		E		F		G		Standard	
Test solutions:	1	2	3	4	5	6	7	8	9	10	11	12	13	14	15	16
	67	20	63	21	62	15	64	15	73	23	73	28	73	30	69	28
	66	21	61	19	62	13	59	18	66	26	71	24	71	31	63	25
	64	17	62	18	59	12	64	19	74	26	67	26	72	28	66	24
	64	18	61	16	57	13	68	12	69	21	66	27	69	26	67	19
Treatment totals	261	76	247	74	240	53	255	64	282	96	277	105	285	115	265	96
Preparation totals	337		321		293		319		378		382		400		361	

[a] The data of example 6e, modified responses.

Table 6.17

Modified Responses Tabulated in the Order They Appeared on the Plate (Example 6e)[a]

	A	B	A	B	B	A	B	A	Row totals	
A	67	64	23	63	28	30	28	62	365	
B	69	73	15	73	15	21	20	73	359	
A	62	61	31	59	25	26	24	66	354	
B	19	13	71	21	66	63	71	18	342	
A	26	24	64	26	62	59	64	28	353	
A	26	27	57	19	68	64	61	21	343	
B	19	17	66	12	72	67	74	18	345	
B	66	69	16	69	18	12	13	67	330	
Column										Grand
totals	354	348	343	342	354	342	355	353	2791	total

[a] Rows and columns are identified thus: A, those containing test solution 1; B, those containing test solution 2.

Regression squares:

$$\frac{(2112 - 679)^2}{64} = \frac{1433^2}{64} = 32{,}085.79$$

Similarly, treatment totals are used to obtain an expression for parallelism squares that must then be modified by row and column corrections:

Parallelism squares:

$$\frac{(261 - 76)^2 + \cdots + (265 - 96)^2}{8} - \left(\begin{matrix} \text{regression} \\ \text{squares} \end{matrix}\right) - \left(\begin{matrix} \text{row} \\ \text{correction} \end{matrix}\right) - \left(\begin{matrix} \text{column} \\ \text{correction} \end{matrix}\right)$$

From the tabulated values for totals of row responses and column responses of the quasi-Latin square design, the row and column corrections are calculated:

Row correction:

$$\frac{[(\text{sum of rows A}) - (\text{sum of rows B})]^2}{64}$$

$$= \frac{[(365 + 354 + 353 + 343) - (359 + 342 + 345 + 350)]^2}{64}$$

$$= (1415 - 1376)^2/64 = 39^2/64 = 23.77$$

Column correction, similarly:

$$(1392 - 1399)^2/64 = (-7)^2/64 = 0.77$$

Parallelism squares is thus:

$$32,155.62 - 32,085.79 - 23.77 - 0.77 = 45.29$$

Components for rows and columns are now calculated as in Example 5e, but with the inclusion of the row and column corrections from above.

Row squares:

$$\frac{365^2 + 359^2 + \cdots + 330^2}{8} - \frac{2791^2}{64} - \text{row correction}$$

$$= 974{,}569/8 - 2791^2/64 - 23.77$$

$$= 121{,}821.12 - 121{,}713.76 - 23.77 = 83.59$$

Column squares:

$$\frac{354^2 + 348^2 + \cdots + 353^2}{8} - \frac{2791^2}{64} - \text{column correction}$$

$$= 973{,}947/8 - 2791^2/64 - 0.77$$

$$= 121{,}743.37 - 121{,}713.76 - 0.77 = 28.84$$

The analysis of variance is summarized in Table 6.18. The deviation from parallelism is shown to be nonsignificant by consideration of the variance ratio:

$$9.06/5.51 = 1.64$$

Table 6.18

Summary of the Analysis of Variance (Example 6e)

Source of variation	Degrees of freedom	Squares	Mean squares
Preparations	7	1,204.86	
Regression (slope)	1	32,085.79	
Parallelism (deviations)	5	45.29	9.06
Row correction	1	23.77	
Column correction	1	0.77	
Subtotal	15	33,360.48	
Treatments (solutions)	15	33,360.49	
Rows (corrected)	6	83.59	
Columns (corrected)	6	28.84	
Error (by difference)	36	198.32	5.51[a]
Total (all sources)	63	33,671.24	

[a] From error mean squares

$$s^2 = 5.51 \quad \text{and} \quad s = (5.51)^{1/2} = 2.34$$

In terms of unmodified observations this represents 0.23 mm.

By reference to the tables for variance ratio (Appendix 10) for $n_1 = 5$ and $n_2 = 36$ it is seen that this value (1.64) lies between the limits set in the 10 and 20% tables, i.e., 2.02 and 1.55, respectively. At such levels the deviations from parallelism are not considered to be significant.

To obtain the confidence limits of the potency estimate first calculate g from the following values using Eq. (6.4):

$$S_{xx} = 32[(0.5)^2 + (-0.5)^2] = 16.00$$
$$b = (2112 - 679)/(32 \times I) = 44.78$$

where $(I = \log 10 = 1.0)$

$$t = 2.02 \quad \text{(for} \quad P = 0.95 \quad \text{and} \quad 36 \quad \text{d.f.)}$$

Thus

$$g = \frac{5.51 \times (2.02)^2}{(44.78)^2 \times 16.00} = 0.0007$$

As g is small, calculate $V(M)$ for Sample A using the approximate Eq. (6.6):

$$V(M_A) = \frac{5.51}{(44.78)^2} \left[\frac{1}{8} + \frac{1}{8} + \frac{(-0.067)^2}{16} \right] = 0.000688$$

Standard error of log potency estimate is obtained by Eq. (6.7) as

$$s_{M_A} = [0.000688]^{1/2} = 0.02623$$

Log percent confidence limits are given by Eq. (6.8) as

$$2 \pm 2.02 \times 0.02623 = 1.9471 \quad \text{to} \quad 2.0529$$

The corresponding percent confidence limits are therefore 88.5–112.9%. Sample potency was estimated in Chapter 2 as 8.6 μg/ml, and so now may be reported as 8.6 μg/ml, limits 7.6–9.7 μg/ml $(P = 0.95)$.

Percent confidence limits for Samples B–G may be calculated in a similar manner and would be almost identical with those for Sample A. The equation of Tootill makes no distinction between samples and gives a value that is applicable to all samples provided that M is not too far from zero.

Applying the appropriate form of Tootill's equation for % standard error,

$$100 \times \ln 10 \times 2 \left[\frac{\text{error mean squares}}{\text{regression squares}} \right]^{1/2}$$

$$= 100 \times 2.3026 \times 2[5.51/32,086]^{1/2} = 6.1\%$$

This corresponds roughly as before to confidence limits of $\pm 12\%$ for $P = 0.95$.

6.11   Evaluation of a Large Plate, Low Precision Assay

The analysis of variance for large plate assays employing an entirely random distribution of test solutions with samples at one dose level only is relatively simple. As test solutions occur in a random manner in both rows and columns it is not possible to isolate a meaningful component for variation between rows and between columns. The procedure is illustrated by the assay of bacitracin, Example 7. It is particularly interesting as it shows the limitations of assay methods when sample potencies are far removed from the range of standard potencies.

Example 7e: Agar Diffusion Assay of Bacitracin by Means of a Low Precision 8 × 8 Random Design Using a Large Plate

For convenience of calculation, responses are modified by first multiplying by 10 and then subtracting 100. These modified responses are shown in Table 6.19.

Analysis of Variance

Total deviation squares (from Eq. (6.1)):

$$S_{yy} = 523,934 - (5186)^2/56$$
$$= 523,934 - 480,260.6 = 43,673.4$$

From the treatment totals in Table 6.19, sums of squares are obtained first for all treatments and then for standard treatments only:

Treatment squares (all):

$$\frac{488^2 + 454^2 + \cdots + 535^2}{4} - \frac{(5186)^2}{56}$$

$$= 523,237.5 - 480,260.6 = 42,976.9$$

Treatment squares (standards only):

$$\frac{289^2 + 371^2 + 444^2 + 535^2}{4} - \frac{(1639)^2}{16}$$

$$= 176,130.75 - 167,895.06 = 8,235.69$$

The standard response line is analyzed using orthogonal polynomial coefficients, as shown in Table 6.20 using the same procedures that were introduced in Example 1e(b).

In the case of a four point dose–response line, this leads naturally to the isolation of components for quadratic and cubic curvature. It is shown in Chapter 8 that for balanced designs, quadratic curvature does not cause bias. In contrast, when the design is unbalanced such as in the this assay, quadratic curvature does bias the estimated potency. Accordingly it seems appropriate to combine the squares for quadratic and cubic curvature to give a pooled mean squares with two degrees of freedom.

Table 6.19

The assay of bacitracin[a]

Test solution number:	1	2	3	4	5	6	7	8	9	10	11	12	Standards potency, IU/ml			
													0.5	1.0	2.0	4.0
													13	14	15	16
	121	110	113	101	43	—	33	—	101	91	92	82	70	91	112	139
	121	115	108	101	49	—	28	—	97	86	90	81	72	93	109	133
	127	108	109	109	39	—	29	—	98	88	88	83	68	95	112	132
	119	121	118	119	42	—	35	—	101	85	84	82	79	92	111	131
Treatment total:	488	454	448	430	173	—	125	—	397	350	354	328	289	371	444	535

[a] The of example 7e, modified responses

Table 6.20

A Part of the Analysis of Variance of Example 7e

	S_4	S_3	S_2	S_1	T_i	e_i	$T_i^2/4e_i$
Regression	+3	+1	−1	−3	811	20	8221.51
Quadratic curvature	+1	−1	−1	+1	9	4	5.06
Cubic curvature	+1	−3	+3	−1	27	20	9.11
Standard treatment total	535	444	371	289			

In this particular example, no matter whether the sources of curvature are pooled or treated separately, the result is the same—curvature is not significant.

It is of interest to note that squares for cubic curvature is greater than squares for quadratic curvature. In an agar diffusion assay, where it is known from long experience that curvature is slight, smooth, and free from inflections, it seems likely that this is an indication of poor practical technique leading to a bias at one or more of the dose levels. It probably does not represent the true nature of the curve that should be obtained in this assay system.

In the summary of the analysis of variance, Table 6.21, it will be seen that the total of the squares for the three components regression, quadratic, and cubic curvature is almost identical with the value obtained as treatment squares for standard only. This is a useful check on arithmetic.

Table 6.21

Summary of the Analysis of Variance (Example 7e)

Source of variation	Degrees of freedom	Squares	Mean squares
Regression	1	8,221.51	
Quadratic curvature	1	5.06	5.06[a]
Cubic curvature	1	9.11	9.11[a]
Subtotal	3	8,235.68	
Treatment squares (standard)	3	8,235.69	
Treatment squares (total)	13	42,976.9	
Residual variation	42	696.5	16.58[b]
Total (all sources)	55	43,673.4	

[a] A pooled value for curvature mean squares with two degrees of freedom is obtained as

$$(5.06 + 9.11)/2 = 7.04$$

[b] From error mean squares

$$s^2 = 16.58 \quad \text{and} \quad s = (16.58)^{1/2} = 4.07$$

In terms of unmodified observations this represents 0.41 mm.

To calculate g first obtain:

$s^2 = 16.58$ from the analysis of variance

$t = 2.02$ for $P = 0.95$ and 42 d.f.

$b = 67.34$ the value from Example 7 modified in conformity with modified responses

$S_{xx} = 1.8120$ obtained as $4[(0.4515)^2 + (0.1505)^2 + (-0.1501)^2 + (-0.4515)^2]$

From Eq. (6.4),

$$g = \frac{16.58 \times (2.02)^2}{(67.34)^2 \times 1.812} = 0.0082$$

As g is small, the variance M of individual samples is obtained from Eq. (6.6). Calculation of confidence limits is shown for two samples only. These are selected so as to demonstrate the influence of M. First, taking Sample 9 (selected because M is relatively small), the term

$$\frac{(M - \bar{x}_S + \bar{x}_T)^2}{S_{xx}}$$

is evaluated as

$$(-0.0475 - 0.1505 + 0.3010)^2/1.812 = 0.005833$$

Second, taking Sample 7 (selected because M is large), the same term is evaluated as

$$(-1.028 - 0.1505 + 0.3010)^2/1.812 = 0.5829$$

Substituting these values in the expression for variance of M, (Eq. (6.6)), for Sample 9,

$$V(M) = \frac{16.58}{(67.34)^2}\left[\frac{1}{4} + \frac{1}{16} + 0.005833\right]$$

$$= 0.003655[0.312,500 + 0.005833] = 0.001163$$

and for Sample 7,

$$V(M) = 0.003655[0.31250 + 0.58290] = 0.003273$$

Confidence limits for the two samples may now be calculated:
  For Sample 9, the standard error of M is obtained by Eq. (6.7) as

$$s_M = [0.001163]^{1/2} = 0.03411$$

which leads by Eq. (6.8) to logarithms of percentage confidence limits of

$$2 \pm 2.02 \times 0.03411 = 1.9210 \quad \text{to} \quad 2.0790 \qquad (P = 0.95)$$

Corresponding confidence limits are 83.4–119.9% of the estimated potency. From Example 7, Chapter 2, potency was estimated at 12.7 IU/ml. Sample potency may be reported therefore as 12.7 IU/ml, limits 10.6–15.2 IU/ml or approximately 13 ± 2 IU/ml $(P = 0.95)$.

For Sample 7, by Eq. (6.7),

$$s_M = [0.003273]^{1/2} = 0.0572$$

which leads by Eq. (6.8) to logarithms of percentage confidence limits of

$$2 \pm 2.02 \times 0.0572 = 1.8845 \quad \text{to} \quad 2.1155 \quad (P = 0.95)$$

Corresponding confidence limits are calculated as 76.5–130.5% of the estimated potency. Sample potency was estimated as 0.133 IU/ml, and so the calculated confidence limits are 0.102–0.173 IU/ml ($P = 0.95$).

However, it would be incorrect to imply such confidence in a report. The sample response lies far outside the standard response range and it is most unlikely that extrapolation of the rectilinear relationship would be valid. Moreover, because the sample as received was believed to be of very low potency, it was applied to the plate without diluting in buffer solution. In this respect the sample differs from the standard.

The correct conclusion to draw from the observation on this sample is that antibiotic activity has been detected and is probably substantially lower than the lowest standard. It was reported appropriately as "less than 0.5 IU/ml."

Normally when potency estimates of low precision are acceptable it is quite unnecessary to quote confidence limits. Accordingly statistical evaluation of such assays is unusual. However, evaluation of this bacitracin assay has served to show what order of precision may be expected and so indicates whether such a design will be acceptable according to varying circumstances.

6.12 Evaluation of an Assay Incorporating Reference Points: The "FDA" Design

The assay described in Section 2.14, in which sample potencies are estimated from a standard curve, differs from all other plate assays illustrated in this work in that each plate cannot be regarded as a self-contained assay. For every sample and each of the standard dose levels numbers 1, 2, 4, and 5 there is a set of three plates. Each plate in the set has three reference doses (standard at dose level number 3) and three other doses of a single test solution, which may be of any sample or one of the other four standards.

In the calculation of potency given in Chapter 2, the reference point used for all responses to a single treatment is the mean of nine reference zones on the three plates in the set for that treatment. While this is a convenient procedure for routine work, a true correction would be based on differences between individual plates, and therefore as reference point for each plate it should use the mean of the three reference responses on that plate. This is the procedure used in this evaluation of Example 8, and to be strictly consistent, new values are calculated in this way for E, b, F, and M.

Evaluation of assays of this design differs from those hitherto described in the following respects:

(1)   The three nonreference responses on each plate are expressed as deviations (positive or negative) from the mean of the three reference responses on that plate. The analysis of variance is then carried out using these deviations in place of actual observations.

(2)   Taking, for example, an assay of only one sample, there is a total of 90 responses (5 sets of 3 plates each having 6 responses). Of these 90 observations, 45 are reference responses and in effect combine via their 15 plate means to form a grand mean with an arbitrarily assigned value of zero. This is because in the calculation of potency, mean responses are corrected with reference to the grand mean, and here we are considering responses as *deviations* from the grand mean. There are in effect then 46 observations, comprising the grand mean reference point and the 45 deviations from it. Thus there are 45 degrees of freedom.

The evaluation procedure applicable to this type of assay design is illustrated by Example 8, the assay of kanamycin.

Example 8e: Agar Diffusion Assay of Kanamycin Using the Five Point Standard Curve Method and Small Plates Each Including Responses to the Reference Dose—"The FDA Method"

The observations of Example 8 are modified by first expressing them as deviations from the appropriate reference point and then multiplying by 100. This is illustrated for the first of the plates for the standard of dose level number 1, that is, 3.2 IU/ml. The mean reference response for this plate is 15.83 mm, obtained as $(16.1 + 15.6 + 15.8)/3 = 15.83$.

The transformation is thus

$$14.60 - 15.83 = -1.23 \rightarrow -123$$
$$14.10 - 15.83 = -1.73 \rightarrow -173$$
$$13.80 - 15.83 = -2.03 \rightarrow -203$$

The values for all responses modified in this way, together with the arbitrary value zero for the reference points, are given in Table 6.22.

Recalculation of Potency

Because now reference points are used for each individual plate, the values of E, F, b, and M are recalculated, leading to a new potency estimate for the sample:

$$E = \tfrac{1}{90}[(2 \times 1350) + (709) - (-660) - (2 \times -1497)] = 78.478$$
$$b = E/\log 1.25 = 78.478/0.069 = 809.9$$
$$M = -19.89/809.9 = -0.0246$$

Table 6.22

Data of Example 8e[a]

Standard dose, IU/ml	1 3.2	2 4.0	Ref 5.0	4 6.25	5 7.81	Sample "5.0"
			Dose levels			
Modified responses	−123	−93		+80	+173	−43
	−173	−53		+100	+143	+7
	−203	−83		+50	+143	−3
	−153	−90		+90	+170	+3
	−193	−70	0	+80	+180	+3
	−163	−40		+50	+160	−27
	−173	−67		+103	+147	−33
	−153	−47		+63	+147	−43
	−163	−117		+93	+87	−43
Totals	−1497	−660	0	+709	+1350	−179

[a] Responses are expressed as 100 times their deviation from the corresponding reference point, which has been assigned an arbitrary value of zero. It follows that the grand mean reference point also has the value zero.

F is obtained simply as the mean of all the nine modified sample responses of Table 6.22 as $-179/9 = -19.89$.

In fact, the newly calculated value for M is identical with the original value found, thus confirming the practical validity of the routine correction procedure.

Analysis of variance.

Total deviation squares (from Eq. (6.1)):

$$S_{yy} = 581{,}817 - (-277)^2/46$$
$$= 581{,}817 - 1668.02 = 580{,}148.98$$

(where -277 is the sum of positive and negative values of all modified responses). From the treatment totals:

Treatment squares:

$$\frac{(-1497)^2 + (-660)^2}{9} + \frac{(0)^2}{1} + \frac{(709)^2 + (1350)^2}{9} - \frac{(-277)^2}{46}$$

$$= 559{,}314.55 - 1{,}668.02 = 557{,}646.54$$

Preparation squares:

$$\frac{(-1497 - 660 + 709 + 1350)^2}{36} + \frac{(0)^2}{1} + \frac{(-179)^2}{9} - \frac{(277)^2}{46}$$

$$= 266.70 + 0 + 3560.11 - 1668.02 = 2158.87$$

As the middose (reference point) has no influence on regression, the squares for regression is calculated using orthogonal polynomial coefficients exactly as if there were an equal number of responses at all five levels, thus:

Regression squares:

$$T_i^2/9e_i$$

where

$$T_i = (-2 \times -1497) + (-660) + (709) + (2 \times 1350) = -7063$$
$$e_i = (-2)^2 + (-1)^2 + (0)^2 + (1)^2 + (2)^2 = 10$$

so that

$$T_i^2/9e_i = -7063^2/90 = 554{,}288.54$$

Squares for deviations from regression plus deviation of mean (nonreference) standards from the reference point, is obtained as the difference

$$\text{(treatment squares)} - \text{(preparation squares)} - \text{(regression squares)}$$
$$= (557{,}646.54) - (2158.87) - (554{,}288.54) = 1199.13$$

This partial analysis of variance is summarized in Table 6.23.

Table 6.23

Summary of the Analysis of Variance (Example 8e)

Source	Degrees of freedom	Squares	Mean squares
Preparations	1	2,158.87	
Regression	1	554,288.54	
Deviations from regression, etc. (by difference)	3	1,199.13	399.70
Treatments	5	557,646.54	
Residual error (by difference)	40	22,502.44	562.56
Total (all sources)	45	580,148.98	

The mean of the pooled value for squares for deviations from regression and deviation of the mean of the nonreference standard responses from the reference point being lower than error mean squares is clearly not significant.

The evaluation may then be continued by first calculating g from

$$s^2 = 562.6$$
$$t = 2.02 \quad \text{for} \quad P = 0.95 \quad \text{and} \quad 40 \quad \text{d.f.}$$
$$b = 809.9$$
$$S_{xx} = 0.8451$$

obtained directly as $9[(0.1938)^2 + (0.0969)^2 + (-0.0969)^2 + (-0.1938)^2]$

Substituting these values in Eq. (6.4),

$$g = \frac{562.6 \times (2.02)^2}{(809.9)^2 \times 0.8451} = 0.0041$$

As g is less than 0.1, the variance of M is calculated by means of the approximate relationship, Eq. (6.6):

$$V(M) = \frac{562.6}{(809.9)^2}\left[\frac{1}{36} + \frac{1}{9} + \frac{(-0.0246)^2}{0.8451}\right]$$

$$= 0.001178[0.139617] = 0.0001198$$

and by Eq. (6.7),

$$s_M = [0.0001198]^{1/2} = 0.010944$$

Logarithms of percentage confidence limits are then obtained by Eq. (6.8) as

$$2 \pm 2.02 \times 0.010944 = 1.9779 \quad \text{to} \quad 2.0221$$

Corresponding confidence limits of the estimated potency are obtained as the antilogarithms and are 95.0–105.2% $(P = 0.95)$.

In Example 8, Chapter 2, it was shown that the value $M = -0.0246$ led to a potency of 767 IU/mg (or 986,000 units of kanamycin activity in a vial of average weight, 1.284 g).

The sample may be reported therefore as having a potency of 767 IU/mg, limits 729–807 IU/mg $(P = 0.95)$.

6.13   Evaluation of a Quantitative Bioautograph

The principle involved in evaluation of a bioautograph in which response is proportional to log dose are identical with those already shown for simple agar diffusion assays. However, the assay used as an example here has some interesting features arising from the imbalance of the design. In the breakdown of variation due to treatments, only the standard response line can exhibit curvature. Deviations from parallelism must be assessed using three dose levels for standard but only two for sample.

The quantitative bioautograph of E129 (Example 18) is used as an illustration.

Example 18e: Bioautograph Assay of E129 by 3 + 2 Design

The observed responses are used without modification, as little simplification of the arithmetic would be achieved by such a step. Actual responses are not reproduced here but can be seen in Example 18, Chapter 5.

Treatment totals for standard high, medium, and low and samples high and low are 642, 425, 181, 545, and 294, respectively. Plate totals are 489, 509, 564, and 525.

Analysis of Variance

Total Deviation Squares (from Eq. (6.1)):

$$S_{yy} = 253,001 - (2087)^2/20 = 253,001.00 - 217,778.45 = 35,222.55$$

Treatment squares:

$$\frac{642^2 + 425^2 + \cdots + 294^2}{4} - \frac{2087^2}{20}$$

$$= 252,252.75 - 217,778.45 = 34,474.30$$

Preparation squares:

$$\frac{(642 + 425 + 181)^2}{12} + \frac{(545 + 294)^2}{8} - \frac{2087^2}{20}$$

$$= 129,792.00 + 87,990.13 - 217,778.45 = 3.68$$

Plate squares:

$$\frac{1,091,923}{5} - \frac{2087^2}{20} = 218,384.60 - 217,778.45 = 606.15$$

The deviation squares due to treatments is broken down into its component parts by means of polynomial coefficients as shown in Table 6.24, using the procedure first introduced in Example 1e(b).

The analysis of variance is summarized in Table 6.25.

Variance ratios for deviations from parallelism and for curvature are 3.55 and 2.57, respectively. Reference to the 5% variance ratio table (Appendix 10) shows the limiting values for $n_1 = 1$ and $n_2 = 12$ to be 4.8. Hence, curvature and deviation from parallelism are considered to be nonsignificant.

Table 6.24

A Part of the Analysis of Variance of Example 18

	Sample		Standard					
	High	Low	High	Medium	Low	e_i	T_i	$T_i^2/4e_i$
Preparations	+3	+3	−2	−2	−2	30	21	3.68
Regression	+1	−1	+2	0	−2	10	1173	34,398.23
Parallelism	+2	−2	−1	0	+1	10	41	42.03
Curvature (of standard)	0	0	+1	−2	+1	6	−27	30.38
Treatment total	545	294	642	425	181			

Table 6.25

Summary of the Analysis of Variance (Example 18)

Sources	Degrees of freedom	Squares	Mean squares
Preparations	1	3.68	
Regression	1	34,398.23	
Parallelism	1	42.03	42.03
Curvature	1	30.38	30.38
Subtotal	4	34,474.32	
Treatments	4	34,474.30	
Plates	3	606.15	
Error	12	142.10	11.84
Total (all sources)	19	35,222.55	

Before calculating confidence limits for the potency estimate, g must first be evaluated. The values of logarithms of doses are 1.3802, 1.0792, and 0.7782 for standard and 1.2553 and 0.9542 for sample. As these are not symmetrically distributed about their combined mean, the simplest calculation of S_{xx} is by Eq. (2.7), thus:

$$S_{xx} = 4(6.1615) - (4 \times 5.4471)^2/20 = 0.9093$$
$$t = 2.18 \quad \text{for} \quad P = 0.95 \quad \text{and} \quad 12 \quad \text{d.f.}$$
$$b = 194.2 \quad \text{(from Example 18)}$$
$$s^2 = 11.84$$

Substituting these values in Eq. (6.4),

$$g = \frac{11.84 \times (2.18)^2}{(194.2)^2 \times 0.9093} = 0.00138$$

As g is less than 0.1, the variance of M is calculated by the simple expression Eq. (6.6):

$$V(M) = \frac{11.84}{(194.2)^2}\left[\frac{1}{12} + \frac{1}{8} + \frac{(-0.0185 - 1.0792 + 1.1048)^2}{0.9093}\right] = 0.0000655$$

The standard error of M is obtained from Eq. (6.7) as

$$s_M = [0.0000655]^{1/2} = 0.0081$$

Logarithms of percent confidence limits are given by Eq. (6.8) as

$$2 \pm 2.18 \times 0.0081 = 1.9821 \quad \text{to} \quad 2.0170 \quad (P = 0.95)$$

The corresponding confidence limits are 96.1–104.0% of the estimated potency.

In Chapter 5, Example 18, potency (G content) was estimated as 345 μg/ml; thus the sample may be reported as containing component G 345 μg/ml, limits 331–358 μg/ml ($P = 0.95$).

6.14   Evaluation of Tube Assays Using Function of Response Versus Logarithm of Dose

It was seen in Chapter 4 that for some antibiotic tube assays a plot of angular transformation of response versus logarithm of dose may approximate to a straight line. There is no defined relationship between this transformation and the mode of action of the antibiotic. It is used in the hope that it will permit evaluation of the assay on the basis of straight parallel log dose–response lines for sample and standard so as to obtain confidence limits of the estimated potency. If this can be done, then ideals in assay design can be formulated, as will be discussed in Chapter 8.

Tube assays normally include several dose levels of standard and so a check on linearity is possible. A well-designed assay will also include more than one dose level of sample, thus permitting a check that response lines are parallel.

If tubes are distributed in racks in such a way that each rack may be regarded as a separate assay, then a component of variance due to racks may be isolated just as a plate component may be isolated in some agar diffusion assays. The tubes should of course be distributed in a random manner, either as a complete set within one rack or throughout all racks. Handling replicate tubes together may disguise the true extent of error, thus leading to a biased potency estimate and a falsely favorable assessment of precision.

The evaluation of this type of assay is illustrated by means of two examples. The first of these (Example 13) was designed for potency estimation by a purely graphical method. Doses do not form a geometrical progression, a fact that makes the computation more tedious. The calculation given in this chapter leads to estimate of the sample potency and its confidence limits. The potency estimate by purely arithmetical means was not shown in Chapter 4.

The observations of Example 13 are transformed to values of ϕ by means of the relationship (Eq. (4.8))

$$p = \sin^2 \phi, \qquad \text{where} \quad p = 100\, A/A_{max}$$

A is the absorbance in an individual tube and A_{max} the mean absorbance of the zero control tubes. The values of ϕ so obtained are multiplied by

10 and then rounded off to three significant figures to facilitate the computation.

The transformation is shown in detail for only the first tube, thus:

The three zero control tubes gave transmittances of 50, 51, and 49%, corresponding to absorbances of 0.301, 0.292, and 0.310. The mean absorbance is then 0.301.

Transmittance for the first standard tube at 4 IU/ml is 58.5%, which is equivalent to an absorbance of 0.233:

$$p = (100 \times 0.233)/0.301 = 77.4\%$$

The value of ϕ corresponding to 77.4% is conveniently obtained from tables as 61.61°. Multiplying by 10 and rounding off to three significant figures this gives the modified value of ϕ as 616.

Such a transformation for all 30 preparation tubes would of course be too tedious in routine work. However, it is quite practicable in an occasional evaluation.

Example 13e: Turbidimetric Assay of Streptomycin in Which Parallel Lines for Each Preparation Are Obtained on Plotting Angular Transformation of Response against Logarithm of Dose

Responses (optical absorbance) are transformed by means of Eq. (4.8) to values of ϕ. Then for convenience of arithmetic, each transformed response is multiplied by 10. The responses modified in this way are shown in Table 6.26.

Although in principle the calculation of sample potency is just the same as in previous examples, because log doses do not form an arithmetical progression the calculation of b is more complex.

Table 6.26

Modified Transformations of Response from Example 13e

Dose, IU/tube		log dose		10 × ϕ		10 × ϕ totals
Standard	4	0.6021	616	625	562	1803
	6	0.7782	482	482	453	1417
	8	0.9031	362	352	384	1098
	10	1.0000	306	280	312	898
	12	1.0792	230	248	256	734
	14	1.1461	230	187	216	633
	16	1.2041	180	129	142	451
Sample	8	0.9031	362	340	380	1082
	10	1.0000	302	291	302	845
	12	1.0792	261	244	244	749

Potency calculation:

$$S_{x_Sx_S} = 20.135,352 - [(20.1384)^2/21]$$
$$= 20.135,352 - 19.312,150 = 0.823,202$$

$$S_{x_Sy_S} = 6138.5627 - [(20.1384)(7034)/21]$$
$$= 6138.5627 - [141,653.5056/21]$$
$$= 6138.5627 - 6745.4050 = -606.8423$$

$$b_S = -606.8423/0.823202 = -737.1732$$

$$S_{x_Tx_T} = 8.940,787 - [(8.9469)^2/9]$$
$$= 8.940,787 - 8.894,113 = 0.046,674$$

$$S_{x_Ty_T} = 2680.4750 - [(8.9469)(2726)/9]$$
$$= 2680.4750 - [24,389.2494/9]$$
$$= 2680.4750 - 2709.9166 = -29.4416$$

$$b_T = -29.4416/0.046,674 = -630.7923$$

Weighted mean slope is given by:

$$b_m = \frac{S_{x_Sy_S} + S_{x_Ty_T}}{S_{x_Sx_S} + S_{x_Tx_T}} \tag{6.9}$$

as

$$\frac{-606.84 - 29.44}{0.8232 + 0.0467} = \frac{-636.28}{0.8699} = -731.47$$

As means of standard dose and nominal sample dose are unequal, it is necessary to calculate estimated potency using Eq. (2.5b). The steps involved are

(1) $\bar{y}_S = 7034/21 = 335.0$
$\bar{y}_T = 2726/9 = 302.9$
$F = \bar{y}_T - \bar{y}_S = 302.9 - 335.0 = -32.06$
$M = F/b_m = -32.06/-731.47 = 0.0438$

(2) $\bar{x}_S = 20.138/21 = 0.9590$
$\bar{x}_T = 8.947/9 = 0.9941$

(3)   Using Eq. (2.5b),

$$M' = 0.0438 + 0.9590 - 0.9941 = 0.0087$$

(4)   Ratio of estimated potency of sample test solution to its nominal potency is given by

$$R' = \text{antilog } M' = \text{antilog } 0.0087 = 1.021$$

By reference to the weighings and dilutions given in Example 13, sample potency is then calculated to be

$$\frac{96.8 \times 1.021 \times 50 \times 50}{5 \times 63.35} = 780 \quad \text{IU/mg}$$

Evaluation

Total deviation squares (from Eq. (6.1)):

$$S_{yy} = 3,656,562 - (9760)^2/30$$
$$= 3,656,562 - 3,175,253 = 481,309$$

Treatment squares:

$$\frac{1803^2 + 1417^2 + \cdots + 749^2}{3} - \frac{(9760)^2}{30}$$

$$= 3,648,767 - 3,175,253 = 473,514$$

Regression squares:

$$b_m(S_{x_S y_S} + S_{x_T y_T}) = -731.465[-606.8423 - 29.4416] = 465,419.5$$

Squares for deviation from parallelism (obtained in a manner analogous to that first used in Example 2e):

$$[b_S S_{x_S y_S} + b_T S_{x_T y_T}] - [\text{regression squares}]$$
$$= [(-737.1732 \times -606.8423) + (-630.7923 \times -29.4416)] - [465,419.5]$$
$$= 500.0$$

Preparation squares:

$$\frac{(7034)^2}{21} + \frac{(2726)^2}{9} - \frac{(9760)^2}{30}$$

$$= 2,356,055 + 825,675 - 3,175,253 = 6477$$

As log doses do not form an arithmetical progression it is not possible to use orthogonal polynomial coefficients to calculate squares for curvature directly. Instead a pooled value is obtained by difference:

$$[\text{total curvature squares}] = [\text{treatment squares}] - [(\text{regression squares})$$
$$+ (\text{parallelism deviation squares})$$
$$+ (\text{preparation squares})]$$
$$= [473514.0] - [(465419.5) + (500.0) + (6477.0)]$$
$$= 1117.5$$

Corresponding degrees of freedom is obtained similarly as:

$$[\text{treatment d.f.}] - [(\text{regression d.f.}) + (\text{parallelism deviations d.f.}) + (\text{preparations d.f.})]$$
$$= (9) - (1 + 1 + 1) = 6$$

As the sample response line has three points only, it contributes no powers of curvature higher than quadratic. Thus the components of the pooled value for curvature are combined quadratic curvature of both standard and sample response lines, the contrast between these two quadratics together accounting for two degrees of freedom. The remaining four degrees of freedom refer to higher powers of curvature of the seven point standard response line.

In fact, even if it were possible to break down the pooled value into its components, no useful purpose would be served in this unbalanced design.

The values for all sums of squares are summarized in Table 6.27.

Reference to variance ratio tables shows that the ratio of mean squares for parallelism : error is not significant. Similarly, the ratio of the pooled value for all degrees of curvature and opposed quadratic curvature is not significant.

Table 6.27

Summary of the Analysis of Variance (Example 13e)

Source of variation	Degrees of freedom	Squares	Mean squares
Regression	1	465,419.5	
Parallelism (deviations)	1	500.0	500.0
Quadratic curvature	1 ⎱		
Opposed quadratic curvature	1 ⎰	1,117.5	186.3
Higher degrees of curvature	4		
Preparations	1	6,477.0	
Treatments	9	473,514.0	
Error	20	7,795.0	389.75
Total (all sources)	29	481,309.0	

To obtain g, the following values are needed:

$$s^2 = 389.75$$
$$t = 2.09 \quad \text{(for} \quad P = 0.95 \quad \text{and} \quad 20 \quad \text{d.f.)}$$
$$b_m = -731.465$$
$$S_{xx} = S_{x_S x_S} + S_{x_T x_T} = 0.823{,}202 + 0.046{,}674 = 0.869{,}876$$

so that

$$g = \frac{389.76 \times (2.09)^2}{(-731.465)^2 \times 0.8699} = 0.0037$$

As g is less than 0.1, variance of M is evaluated by Eq. (6.6) as

$$V(M) = \frac{389.76}{535{,}041} \left[\frac{1}{21} + \frac{1}{9} + \frac{(0.00870 - 0.95897 + 0.99410)^2}{0.869{,}876} \right] = 0.0001172$$

Then by Eq. (6.7),

$$s_M = [0.0001172]^{1/2} = 0.01083$$

Log of percentage confidence limits are given by Eq. (6.8) as

$$2 \pm 2.09 \times 0.01083 = 1.9774 \text{ to } 2.0226$$

Corresponding confidence limits are 94.9–105.3% of the estimated potency.

In example 13, Chapter 4, potency was calculated as 780 IU/mg. Sample potency may be reported as 780 IU/mg (limits 740–823, $P = 0.95$).

The advantage of purpose built designs is again demonstrated by the relative simplicity of the evaluation of a balanced tube assay using doses forming a geometrical progression. As explained in Chapter 4, the assay of tetracycline (Example 15) was designed so that responses to extreme doses could be discarded if necessary. Inspection of the response curve (Fig. 4.6d) shows that responses to the four central doses approximate to a straight line, whereas if all six response levels are considered the line tends to be slightly sigmoid. Accordingly this assay was evaluated on the basis of the four central dose levels only.

Responses were expressed in terms of 10ϕ as in Example 13e, using tables based on the relationship Eq. (4.8), $p = \sin^2 \phi$.

Example 15e: Balanced Turbidimetric Assay of Tetracycline in Which Parallel Lines Are Obtained on Plotting Angular Transformation of Response against Logarithm of Dose for Both Preparations

Responses (optical absorbance) are transformed by means of Eq. (4.8) to values of ϕ. Then for convenience of arithmetic, each transformed response is multiplied by 10.

The responses modified in this way are shown in Table 6.28.

Table 6.28

Modified Transformations of Response from Example 15e

Dose, IU/tube		log dose	$10 \times \phi$			$10 \times \phi$ totals
Standard	0.120	$\bar{1}.0792$	512	530	512	1554
	0.180	$\bar{1}.2553$	439	443	439	1321
	0.270	$\bar{1}.4314$	348	348	348	1044
	0.405	$\bar{1}.6075$	271	271	271	813
						4732
Sample	0.120	$\bar{1}.0792$	564	556	556	1676
	0.180	$\bar{1}.2553$	477	477	477	1431
	0.270	$\bar{1}.4314$	387	382	382	1151
	0.405	$\bar{1}.6075$	295	295	291	881
						5139

Analysis of variance

Total deviation squares (from Eq. (6.1)):

$$S_{yy} = 4,289,817 - 9871^2/24$$
$$= 4,289,817 - 4,059,860 = 229,957$$

Treatment squares:

$$\frac{1554^2 + 1321^2 + \cdots + 881^2}{3} - \frac{9871^2}{24} = 4,289,520 - 4,059,860 = 229,660$$

Preparation squares:

$$\frac{4732^2 + 5139^2}{12} - \frac{9871^2}{24} = 4,066,762 - 4,059,860 = 6902$$

The treatment squares is further broken down into its components by means of polynomial coefficients, as shown in Table 6.29.

Table 6.29

Part of the Analysis of Variance of Example 15e

	Sample				Standard				e_i	T_i	$T_i^2/3e_i$
	x_{T_5}	x_{T_4}	x_{T_3}	x_{T_2}	x_{S_5}	x_{S_4}	x_{S_3}	x_{S_2}			
Preparations	+1	+1	+1	+1	−1	−1	−1	−1	8	407	6,902.04
Regression	+3	+1	−1	−3	+3	+1	−1	−3	40	−5165	222,310.20
Parallelism	+3	+1	−1	−3	−3	−1	+1	+3	40	−165	226.88
Curvature											
quadratic	+1	−1	−1	+1	+1	−1	−1	+1	8	−23	22.04
opposed quadratic	+1	−1	−1	+1	−1	+1	+1	−1	8	−27	30.38
cubic	+1	−3	+3	−1	+1	−3	+3	−1	40	+135	151.88
opposed cubic	+1	−3	+3	−1	−1	+3	−3	+1	40	−45	16.88
Treatment											
total:	881	1151	1431	1676	813	1044	1321	1554			

The analysis of variance is summarized in Table 6.30 from which it is seen that mean squares for both parallelism deviations and cubic curvature are much greater than error mean squares. Reference to variance ratio tables (e.g., in Appendix 10) shows that for $n_1 = 1$ and $n_2 = 16$, $f = 4.5$ (5% table) and 8.5 (1% table).

It is therefore necessary to consider what this might mean in practical terms. First the physical significance of the error term must be explained.

A value of 18.56 for s^2 corresponds to

$$s = (18.56)^{1/2} = 4.31$$

which in turn leads to a value of 0.431 for the standard error of ϕ. (All values of ϕ had

Table 6.30

Summary of the Analysis of Variance (Example 15e)

Source of variation	Degrees of freedom	Squares	Mean squares
Preparations	1	6,902.04	
Regression	1	222,310.20	
Parallelism	1	226.88	226.88
Curvature (quadratic)	1	22.04	22.04
Opposed quadratic curvature	1	30.38	30.38
Curvature (cubic)	1	151.88	151.88
Opposed cubic curvature	1	16.88	16.88
Subtotal	7	229,660.30	
Treatments	7	229,660	
Error by difference	16	297	18.56
Total (all sources)	23	229,957	

been multiplied by 10 for convenience of computation.) The change in absorbance corresponding to a change in ϕ of 0.431 is not a constant but varies with the value of A/A_{max} from which ϕ was derived. It can be seen by reference to tables of angular transformation (Appendix 8) that in this example, the standard deviation of absorbance corresponding to a standard deviation of $\phi = 0.431$ varies from about 0.0024 (for dose range 0.08–0.12 IU/tube) to about 0.0019 (for dose range 0.405–0.6075 IU/tube). These values of standard deviation seem very small for the manual measurement of absorbance of different cell suspensions of (ideally) the same cell concentration.

On inspection of the replicate observations (Table 4.4), one is bound to wonder whether, having seen the first of each set of three readings, the operator tended to anticipate the second and third readings rather than measure absorbance of the suspensions objectively. In this way, errors in the first reading of each set could lead to apparent differences in slope or apparent curvature. At the same time the apparently high reproducability of observations would lead to low residual error with consequent high significance of the parallelism (deviations) and curvature mean squares.

The above explanation is plausible but unproven. To form any valid conclusion it would be necessary to repeat the assay and evaluation several times with particular attention to technique of optical measurements.

The best that can be done with the existing results is to assume that this explanation is correct and calculate confidence limits by incorporating all deviations from regression in a pooled value for residual error.

This may be calculated as

[total deviation squares] − [(preparation squares) + (regression squares)]
= [229957] − [6902 + 222310] = 745

The corresponding degrees of freedom are

$$[23] - [1 + 1] = 21$$

Thus, pooled error mean squares is $745/21 = 35.5$.

To obtain confidence limits first calculate the following values:

$$s^2 = 35.5 \quad \text{(from the analysis of variance)}$$
$$t = 2.09 \quad \text{(for} \quad P = 0.95 \quad \text{and} \quad 21 \quad \text{d.f.)}$$
$$b = 488.9 \quad \text{(from Example 15, as } 10 \times 48.89)$$
$$S_{xx} = 0.930336$$

S_{xx} was calculated as

$$6[0.26415)^2 + (0.08805)^2 + (-0.08805)^2 + (-0.26415)^2]$$

Then by Eq. (6.4),

$$g = \frac{35.5 \times (2.09)^2}{(488.9)^2 \times 0.9303} = 0.00073$$

As g is less than 0.1, variance of M is evaluated by Eq. (6.6) as

$$V(M) = \frac{35.5}{239000}\left[\frac{1}{12} + \frac{1}{12} + \frac{(-0.0675)^2}{0.9303}\right] = 0.0000268$$

Then by Eq. (6.7),

$$s_M = [0.0000268]^{1/2} = 0.00517$$

Log of percent confidence limits are then given by Eq. (6.8) as

$$2 \pm 2.09 \times 0.00517 = 1.9892 \quad \text{to} \quad 2.0108$$

Corresponding confidence limits are 97.5–102.5%. Sample potency was estimated as 21.4 mg/ml (Example 15, Chapter 4) so that it may be reported as 21.4 mg/ml, limits 20.9–21.9 mg/ml ($P = 0.95$).

References

Emmens, C. W. (1948). "Principles of Biological Assay." Chapman and Hall, London.
Finney, D. J. (1964). "Statistical Method in Biological Assay." Griffin, London.
Fisher, R. A., and Yates, F. (1963) "Statistical Tables for use in Biological, Agricultural and Medical Research" 6th ed. Longman, London.
Lees, K. A., and Tootill, J. P. R. (1955). *Analyst* **80**, 112.

CHAPTER 7

EVALUATION OF SLOPE RATIO ASSAYS

7.1 Principles of Evaluation

The evaluation of slope ratio assays is in general principle the same as for parallel line assays. Response variation due to all known causes is isolated, thus leading to a residual variation corresponding to the random error. From this residual variation an estimate of random error of the determined potency may be computed.

However, the technique of evaluation differs from that for parallel line assays in several respects.

(1)   For each preparation there is an independent regression coefficient and so regression squares has a number of degrees of freedom equal to the number of preparations.

(2)   In place of "deviations from parallelism" there are the criteria of validity: "failure of regression lines to coincide at zero dose" and "discrepancy of the lines from response to zero dose control." The calculations of squares corresponding to these discrepancies and also squares for curvature is relatively tedious. However, a pooled mean square value corresponding to all these sources is often an adequate validity check. Such a simplified version of the analysis of variance serves to illustrate principles of assay design. Accordingly, in the examples given here all such deviations from the ideal are grouped together as "deviations from regression." For a fuller study of deviations from regression the reader may consult Finney (1964) or Bliss (1951).

(3)   Total regression squares is obtained as the sum of the products of b and S_{xy} for all preparations, thus:

$$\text{Regression squares} = b_S S_{x_S y} + b_T S_{x_T y}, \quad \text{etc.} \quad (7.1)$$

(4)   The g criterion for significance of the standard slope is given by

$$g = \frac{s^2 \times t^2 \times S_{x_T x_T}}{b_S^2 [(S_{x_S x_S})(S_{x_T x_T}) - (S_{x_S x_T})^2]} \quad (7.2)$$

(5)   Variance of the potency ratio is given by a lengthy "exact" expression that incorporates g. As its seems that in microbiological slope ratio assays g is invariably small, the full expression is not given here. When g is less

than 0.1, the following simplified expression for $V(R)$ is a very good approximation:

$$V(R) = \frac{s^2}{b_S{}^2} \cdot \frac{S_{x_Sx_S} + 2RS_{x_Sx_T} + R^2S_{x_Tx_T}}{(S_{x_Sx_S})(S_{x_Tx_T}) - (S_{x_Sx_T})^2} \tag{7.3}$$

Equations (7.1–7.3) are those of Finney (1950).

7.2   Evaluation of Simple Slope Ratio Assays

These are illustrated by reference to the two assays for pantothenic acid, Examples 10 and 11, Chapter 3. The observations, arbitrary units as measured by EEL nephelometer, were used directly in the calculation without modification, and so they are not repeated here.

Example *10e*:   Assay of Pantothenic Acid—an Unbalanced Design

Total deviation squares (from Eq. (6.1)):

$$S_{yy} = 35,358 - 760^2/20 = 35,358 - 28,880 = 6478$$

Regression squares (from Eq. (7.1)):

$$b_S S_{x_Sy} + b_T S_{x_Ty}$$

Substituting the values from Example 10, this becomes

$$(18.349 \times 208) + (26.955 \times 97) = 6431.23$$

Treatment squares:

$$\frac{83^2 + 166^2 + 235^2}{4} + \frac{95^2 + 172^2}{3} + \frac{9^2}{2} - \frac{760^2}{20}$$

$$= 22,417.50 + 12,869.67 + 40.5 - 28,880.00 = 6447.67$$

All 20 tubes were placed in the same rack in a randomized manner, so that no further isolation of effects was possible. The squares are summarized in Table 7.1 and a pooled value for deviations from regression with three degrees of freedom is obtained by

(treatment squares) − (regression squares) = 6447.67 − 6431.23 = 16.44.

Mean squares for deviation from regression are compared with error mean squares and shown to be nonsignificant; thus,

$$5.48/2.17 = 2.53$$

Reference to the 5% variance ratio table shows that the limiting value for $n_1 = 3$ and $n_2 = 14$ is 8.7. The value 2.53 being much less than 8.7, deviations from regression are not considered to be significant.

Table 7.1

Summary of Analysis of Variance (Example 10e)

	Source of variation	Degrees of freedom	Squares	Mean squares
1.	Regression	2	6431.23	
2.	Deviations from regression (by difference)	3	16.44	5.48
3.	Treatments	5	6447.67	
4.	Error (by difference)	14	30.33	2.17
5.	Total	19	6478.00	
	Pooled error (2 + 4)	17	46.77	2.75

As in fact the deviations from regression appear to be part of the random error, it seems reasonable to examine the effect of combining these squares with the residual error. A value of $30.33 + 16.44 = 46.77$ with 17 degrees of freedom is obtained. This leads to a mean square value of 2.75. Using this value instead of 2.17, slightly wider but probably more realistic confidence limits are obtained.

To calculate $V(R)$ and also g, the following values are needed:

$$s^2 = 2.75 \quad \text{the total pooled residual error mean square}$$

$$t = 2.11 \quad \text{the value corresponding to 17 d.f., } P = 0.95$$

$$\left.\begin{array}{l} b_S = 18.35 \\ S_{x_S x_S} = 27.20 \\ S_{x_T x_T} = 10.95 \\ S_{x_S x_T} = -10.80 \\ R = 1.469 \end{array}\right\} \quad \text{from Example 10, Section 3.7.}$$

Substituting in Eq. (7.2),

$$g = \frac{2.75 \times (2.11)^2 \times 10.95}{(18.35)^2[(27.20)(10.95) - (-10.80)^2]} = 0.002$$

As g is small, calculate variance of R using Eq. (7.3):

$$V(R) = \frac{2.75}{(18.35)^2}\left[\frac{(27.20) + (2 \times 1.47 \times -10.80) + (1.47^2 \times 10.95)}{(27.20)(10.95) - (-10.80)^2}\right]$$

$$= 0.000861$$

The standard error of R, s_R, is obtained by

$$s_R = [V(R)]^{1/2} \tag{7.4}$$

which in this case is

$$[0.0008612]^{1/2} = 0.02935$$

Confidence limits of R are then given by

$$R \pm t s_R \tag{7.5}$$

Substituting the appropriate values, limits are

$$1.469 \pm 2.11 \times 0.02935 = 1.469 \pm 0.062$$

or 1.407 to 1.531 ($P = 0.95$), corresponding to a $\pm 4.2\%$ range from the estimated potency.

The sample potency was thus estimated as 9.2 mg of calcium pantothenate per tablet, limits 8.8–9.6 mg ($P = 0.95$).

Example 11e: Assay of Pantothenic Acid—a Balanced Design

Total deviation squares (from Eq. (6.1)):

$$S_{yy} = 55,110 - 1082^2/28 = 55,110.00 - 41,811.57 = 13,298.43$$

Regression squares (from Eq. (7.1)):

$$b_S S_{xSy} + b_T S_{xTy}$$

Substituting the values from Example 11, we obtain

$$(17.525 \times 143.572) + (23.204 \times 461.572) = 13,226.42$$

In contrast to Example 10e, as all treatments have the same replication, we have:

Treatment squares:

$$\frac{83^2 + 158^2 + \cdots + 14^2}{4} - \frac{1082^2}{28} = 55,046.50 - 41,811.57 = 13,234.93$$

In this case it would be relatively simple (as compared with Example 10e) to determine separately squares for curvature. However, it is clear by inspection of the graph, Fig. 3.9, that response lines are straight, and so a pooled value for deviations from regression is considered adequate. It is obtained as

$$\begin{bmatrix} \text{squares for deviations} \\ \text{from regression} \end{bmatrix} = \begin{bmatrix} \text{squares for} \\ \text{treatments} \end{bmatrix} - \begin{bmatrix} \text{regression} \\ \text{squares} \end{bmatrix}$$

It has four degrees of freedom.

Sums of squares are summarized in Table 7.2.

The mean squares value for deviations from regression is smaller than the error mean squares, and so without reference to tables the former is clearly not significant. As in the case of Example 10e, the best estimate of residual error mean squares is the pooled value for residual error plus deviations from regression. This leads to an error mean squares value of 2.88 with 25 degrees of freedom.

In order to evaluate variance of R (and also g) the following values are needed:

$$s^2 = 2.88 \qquad \text{the total pooled residual error}$$

$$t = 2.06 \qquad \text{the value for 25 d.f., } P = 0.95$$

Table 7.2

Summary of Analysis of Variance (Example 11e)

	Source of variation	Degrees of freedom	Squares	Mean squares
1.	Regression	2	13,226.42	
2.	Deviations from regression	4	8.51	2.13
3.	Treatments	6	13,234.93	
4.	Error (by difference)	21	63.50	3.02
5.	Total	27	13,298.43	
	Pooled error, (2 + 4)	25	72.01	2.88

$$
\left.
\begin{aligned}
b_S &= 17.525 \\
S_{x_S x_S} &= 35.429 \\
S_{x_T x_T} &= 35.429 \\
S_{x_S x_T} &= -20.571 \\
R &= 1.324
\end{aligned}
\right\} \quad \text{from Example 11, Section 3.8}
$$

Substituting these values in Eq. (7.2), it is found that g is 0.0017 and so $V(R)$ is calculated using the approximate expression, Eq. (7.3):

$$
V(R) = \frac{2.88}{(17.53)^2} \left[\frac{(35.43) + (2 \times 1.324 \times -20.57) + (1.324^2 \times 35.43)}{(35.43)(35.43) - (-20.57)^2} \right]
$$

$$
= 0.0004854
$$

The standard error of R, s_R, is obtained by Eq. (7.4) as

$$
[0.0004854]^{1/2} = 0.002203
$$

Confidence limits of R are then given by Eq. (7.5) as

$$
1.324 \pm 2.06 \times 0.2203 = 1.279 \quad \text{to} \quad 1.369 \qquad (P = 0.95)
$$

These limits correspond to a $\pm 3.4\%$ range from the estimated potency.

The sample potency was thus estimated as 5.47 mg of calcium pantothenate per tablet, limits 5.28–5.65 mg ($P = 0.95$).

7.3 Evaluation of a Multiple Slope Ratio Assay

The simultaneous assay of three preparations of folic acid, Example 12, Chapter 3, is used to illustrate the procedure. The calculation differs only little from the two previous examples. Regression squares is obtained as the sum of four components. As this is a balanced design, values of $S_{x_T x_T}$ are the same for all three samples and so there is only one value of g. The observations of Example 12 are not modified and so are not repeated here.

Example 12e: Assay of Folic acid—a Multiple Balanced Design

Total deviation squares (from Eq. (6.1) and omitting responses to zero dose):

$$S_{yy} = 96{,}504.00 - 1634^2/32 = 96{,}504.00 - 83{,}436.12 = 13{,}067.88$$

In the original potency calculation (Chapter 3) the simplified method using Bliss' formula was employed and so values of S_{xy} were not obtained. It is necessary to calculate these four values now. Thus for the standard,

$$S_{x_S y} = S(x_S y) - (Sx_S)(Sy)/n = 674 - (12)(1634)/32 = 674.00 - 612.75 = 61.25$$

The corresponding values for the three samples A, B, and C are obtained similarly:

$$S_{x_A y} = 745.00 - 612.75 = 132.25$$
$$S_{x_B y} = 646.00 - 612.75 = 33.25$$
$$S_{x_C y} = 705.00 - 612.75 = 92.25$$

Regression squares is then obtained by substitution of these four values and the corresponding values of b (from Example 12, Chapter 3) in a form of Eq. (7.1).

Regression squares:

$$b_S S_{x_S y} + b_A S_{x_A y} + b_B S_{x_B y} + b_C S_{x_C y}$$
$$= (38.95 \times 61.25) + (42.50 \times 132.25) + (37.55 \times 33.25) + (40.50 \times 92.25)$$
$$= 12{,}990.98$$

Treatment squares:

$$\frac{122^2 + 276^2 + \cdots + 290^2}{4} - \frac{1634^2}{32} = 96{,}432.50 - 83{,}436.12 = 12{,}996.38$$

A pooled value for deviations from regression with three degrees of freedom is obtained as before. The analysis of variance is summarized in Table 7.3.

Table 7.3

Summary of Analysis of Variance (Example 12e)

Source of variation	Degrees of freedom	Squares	Mean squares
1. Regression	4	12,990.98	
2. Deviations from regression	3	5.40	1.80
3. Treatments	7	12,996.38	
4. Error	24	71.50	2.98
5. Total	31	13,067.88	
Pooled error, (2 + 4)	27	76.90	2.85

In this case, too, mean squares for deviations from regression is clearly not significant when compared with the error mean squares value as originally calculated. These two sources of error are therefore combined to give a pooled value for error mean squares of 2.85 with 27 degrees of freedom.

To evaluate $V(R)$ the following values are obtained:

$$s^2 = 2.85 \qquad \text{the total pooled residual error}$$
$$t = 2.07 \qquad \text{for 27 d.f. and } P = 0.95$$
$$b_S = 38.95 \qquad \text{from Example 12}$$
$$S_{x_S x_S} = 15.5 \qquad \text{from } [4(1^2 + 2^2)] - [(4 \times 3)^2/32]$$
$$S_{x_T x_T} = 15.5 \qquad \text{as for } S_{x_S x_S}$$
$$S_{x_S x_T} = -4.5 \qquad \text{from } [0] - [(12)(12)/32]$$

Substituting these values in Eq. (7.2), g is obtained as 0.0006. As g is small, the variance of each potency ratio is calculated separately, using Eq. (7.3) for each of the unknown preparations.

For Sample A,

$$V(R_A) = \frac{2.85}{(38.95)^2} \left[\frac{(15.50) + (2 \times 1.091 \times -4.50) + (1.091^2 \times 15.50)}{(15.50)(15.50) - (-4.50)^2} \right]$$

$$= 0.0002059$$

Then by Eq. (7.4),

$$s_{R_A} = (0.0002059)^{1/2} = 0.01435$$

Confidence limits are obtained by Eq. (7.5) as

$$1.091 \pm 2.07 \times 0.01435 = 1.091,$$
$$\text{limits} \quad 1.061 \quad \text{to} \quad 1.121 \quad \text{at} \quad P = 0.95$$

Taking into consideration the weighings and dilutions shown in Example 12, Chapter 3, the sample potency may be reported as 110 μg (limits 107–113 μg) of folic acid per tablet of average weight 443 mg ($P = 0.95$).

For Samples B and C

$$V(R_B) = 0.0001811, \qquad s_{R_B} = 0.01346$$

and confidence limits are

$$0.964 \pm 2.07 \times 0.01346 = 0.964,$$
$$\text{limits} \quad 0.936 \quad \text{to} \quad 0.992 \quad \text{at} \quad P = 0.95$$

Potency of sample B may be reported as 103 μg (limits 100–106 μg) of folic acid per tablet of average weight 452 mg ($P = 0.95$), and,

$$V(R_C) = 0.0001954, \qquad s_{R_C} = 0.01398$$

and confidence limits are

$$1.040 \pm 2.07 \times 0.01398 = 1.040,$$

$$\text{limits} \quad 1.011 \quad \text{to} \quad 1.069 \quad \text{at} \quad P = 0.95$$

Potency of sample C may be reported as 437 μg (limits 424–448 μg) of folic acid per tablet of average weight 103 mg ($P = 0.95$).

7.4 Evaluation of a Turbidimetric Antibiotic Assay by Angular Transformation of Response and Slope Ratio

If any transformation of the response when plotted against dose gives a straight line, then it seems likely that the slope ratio method of potency estimation should be applicable. Such is the case in Example 14, Chapter 4. It has been demonstrated graphically (Fig. 4.5f) that an angular transformation of response gives a linear plot against doses up to 12 IU/tube. It also appears that extrapolated response lines converge toward a single point as zero dose is approached. However, this point does not correspond with the observed response to zero dose. This discrepancy does not invalidate the basis of the calculation (see (3.5)). Potency ratio is calculated omitting the observed responses to zero dose as well as doses 14 and 16 IU/tube.

To simplify the computation, values of ϕ corresponding to responses are expressed to only one decimal place and then multiplied by ten.

Calculation of a single modified response is illustrated by the first response to dose 4 IU/tube of standard. Thus: Observed responses are transmittance at 530 nm. Mean transmittance of the three zero control tubes is 38%, corresponding to a mean absorbance of 0.420. Transmittance of the first tube of standard dose at 4 IU/tube is 45.5%, corresponding to an absorbance of 0.342. This can be obtained as log 100 − log 45.5 = 0.342, or by reference to tables, or better still by reading directly as absorbance from the instrument, provided of course that the instrument is so calibrated that absorbance can be read easily from the scale. Expressed as a percentage relative to absorbance for the zero control tube, this figure becomes $100 \times 0.342/0.420 = 81.4\%$.

Using Eq. (4.8) or more conveniently by obtaining values of ϕ from tables such as shown in Appendix 8, it is seen that for $p = 81.4\%$, $\phi = 64.45°$. Rounding off to one decimal place and multiplying by 10, this becomes 645, which is a more convenient figure for purposes of statistical evaluation.

This whole procedure applied to each observed response is less tedious than it may appear from the description.

In Chapter 4, potency was not calculated via the angular transformation of response and so it must be done here as part of the evaluation.

Example 14e: Turbidimetric Assay of Streptomycin,—an Unbalanced Design

The transformed and modified responses are shown in Table 7.4.

Table 7.4

Modified Transformations of Response (Example 14e)

Dose, IU/tube		Modified responses, 10ϕ				Totals
Standard	4	645	672	613	607	2537
	6	530	514	519	508	2071
	8	447	442	451	442	1782
	10	321	317	335	331	1304
	12	252	242	237	243	974
Sample	8	621	621	592	592	2426
	10	553	572	565	536	2226
	12	548	491	530	530	2099
Zero control	0	900 (by definition)[a]				

[a] Zero control corresponds to $p = 100\%$ and to a ϕ value of $90°$.

Calculation of potency, excluding response to zero dose, $n = 32$:

$S_{x_Sx_S} = [4(4^2 + \cdots + 12^2)] - [(4 \times 40)^2/32] = 640$

$S_{x_Tx_T} = [4(8^2 + \cdots + 12^2)] - [(4 \times 30)^2/32] = 782$

$S_{x_Sx_T} = (0) - [(4 \times 40)(4 \times 30)/32] = -600$

$S_{x_Sy} = [(4 \times 2537) + \cdots + (12 \times 974)] - [(160)(15,419)/32] = -15,537$

$S_{x_Ty} = [(8 \times 2426) + (10 \times 2226) + (12 \times 2099)] - [(120)(15,419)/32] = 9035$

Substituting the values so obtained in the matrix equations (3.10a and b) b_S and b_T are obtained directly as

$$b_S = \frac{(-15537)(782) - (9035)(-600)}{(640)(782) - (-600)} = -47.90$$

$$b_T = \frac{(9035)(640) - (-15537)(-600)}{(640)(782) - (-600)} = -25.20$$

Potency ratio (from Eq. (3.5)):

$$R = -25.20/-47.90 = 0.5261$$

Analysis of variance

Total deviation squares (from Eq. (6.1)):

$$7,955,795 - 15,419^2/32 = 526,249$$

Regression squares (from Eq. (7.1)):

$$(-47.90 \times -15,537) + (-25.20 \times 9035) = 516,563$$

Treatment squares:

$$\frac{2537^2 + 2071^2 + \cdots + 2099^2}{4} - \frac{15,419^2}{32} = 519,546$$

Squares for deviations from regression is found as the difference between treatment squares and regression squares. The analysis of variance is summarized in Table 7.5.

Table 7.5

Summary of Analysis of Variance (Example 14e)

Source	Degrees of freedom	Squares	Mean squares
Regression	2	516,563	
Deviations from regression (by difference)	5	2,983	597
Treatments	7	519,546	
Residual error (by difference)	24	6,703	279
Total	31	526,249	

The variance ratio, regression deviation mean squares to residual error mean squares, is $597/279 = 2.14$. Reference to variance ratio tables shows the limiting values for $n_1 = 5$ and $n_2 = 24$ to be 2.62 (5% table) and 2.09 (10% table), and so although the value 2.14 is rather high there is not *strong* evidence that the deviations from regression are attributable to anything but random error. In these circumstances it seems reasonable to use the value 279 for error mean squares in the calculation of g and $V(R)$ and not calculate a pooled value as in Examples 12e and 13e. Whichever course were adopted the result would be virtually the same in this example.

The g criterion is calculated by substituting in Eq. (7.2) the value $t = 2.064$ (for 24 d.f., $P = 0.95$) and the other appropriate values already found in this example. It is found that g is 0.0029, and so the variance of R is calculated from Eq. (7.3) as

$$V(R) = \frac{279}{(47.90)^2} \left[\frac{(640) + (2 \times 0.526 \times -600) + (0.526^2 \times 782)}{(640)(782) - (-600)^2} \right] = 0.0001949$$

Then by Eq. (7.4),

$$s_R = (0.0001949)^{1/2} = 0.01396$$

Confidence limits are then given by Eq. (7.5) as

$$0.526 \pm 2.064 \times 0.01396$$

or

$$0.497 \quad \text{to} \quad 0.555 \qquad \text{at} \quad P = 0.95$$

Taking into consideration weighings and dilutions given in Example 14, Chapter 4, the sample potency may be reported as 495 IU/mg, limits 468–523 IU/mg ($P = 0.95$).

References

Bliss, C. I. (1951). *In* "The Vitamins" (P. Gyorgy, ed.), Vol. II. Academic Press, New York.

Finney, D. J. (1950). *In* "Biological Standardisation" (J. H. Burn, ed.). Oxford Medical Publ. Oxford.

Finney, D. J. (1964). "Statistical Method in Biological Assay." Griffin, London.

CHAPTER 8

CHOICE OF METHOD AND DESIGN

8.1 Choice of Method

The two most widely used manual procedures, that is, plate and tube assays, both have certain advantages and disadvantages. Both procedures have their protagonists and probably choice is often based on the analyst's personal preference, which is related to his experience and differing degrees of success with the two techniques.

For the assay of some substances the analyst will have little option. For example, it appears that due to its poor diffusion rate, no satisfactory plate method has been devised for tyrothricin. The tube method for cephaloridine is stated to be unsuitable for samples containing decomposition products (Simmons, 1972). If samples are turbid even after dilution to assay level, then a tube (turbidimetric) assay is not applicable.

Setting aside personal preferences and experience, the indisputable facts for antibiotic assays may be summarized.

(1) Tube assays require an incubation period of only about four hours and so results may be obtained the same day that tests are set up. Plate assays normally require overnight incubation.

(2) Tube assays generally have a narrow working range, overall dose ratio being less than fivefold in most assays. In contrast, plate assays may have a much wider usable working range. The ratio of highest to lowest practicable doses may approach 100:1. Within this wide range, however, approximately linear relationships between zone diameter and log dose would normally apply over at least a 2:1 ratio or perhaps even an 8:1 ratio. In Chapter 2, Examples 1 and 7 show response relationships that are approximately linear over an 8:1 dose range.

(3) Tube assays measure total antibiotic activity. Plate assays, being dependent on diffusion rates of the active substance, may not measure total activity. When the antibiotic consists of a mixture of active compounds, then those components having lower diffusion rates may not play any part in fixing the position of the zone boundary even though they are microbiologically active. If the proportions of the various active substances in sample and standard are dissimilar, then assay by diffusion method may yield results quite different from those obtained by the tube method. Garrett 'et al. (1971)

refer to such an occurrence in the assay of deteriorated samples of tetra-cycline, which gave falsely high potencies. They attribute this increase to the presence of anhydrotetracycline. According to Garrett, anhydrotetracycline shows twice the potency of tetracycline by the plate method. He suggests that this might be due to its having greater diffusibility and thus giving a larger inhibition zone despite having a lower true potency. The tube assay being uninfluenced by diffusibility would give a more realisitic estimate of potency.

Other points of contrast between these two basic methods, this time re-flecting the combined or differing experiences of many analysts, include the following:

(1)   The plate method has been preferred in Britain for 20 years or more as being more economical in manpower. Evidence from the U.S. FDA laboratories (Garth, 1972) suggests that the tube method is more economical in manpower. Output per technician working in a team appeared to be about double that achieved by teams working with the plate method. Nevertheless, the plate method is still used there for a substantial proportion of all samples.

(2)   Reproduceability of assay conditions: In many laboratories it is found difficult to control tube assay conditions so that similar response lines and median responses are reproduced from day to day. While such changes affect sample and standard equally and so do not lead to any bias, they can be inconvenient, possibly necessitating the discarding of observations at one extreme or other of the dose levels and thus precluding the application of a routine computation procedure. This is probably a factor contributing to the popularity of graphical methods for estimation of potency.

In contrast, test conditions in the plate method may be controlled so as to yield adequately reproducible responses from day to day. Thus a simple standard routine form of potency computation may be employed.

Considering now the vitamin assays, tube methods are generally very much more sensitive than plate assays and seem to be more commonly used than the former. Tube assays are of particular value in the estimation of vitamins in natural products where concentrations may be very low. Incubation periods are normally at least 16 hours and may be very much longer, so that in this respect they have no advantage over plate methods.

For the examination of pharmaceutical products where concentrations are high, the less sensitive plate methods have the advantage that less dilution of sample and standard is necessary.

However, of paramount importance in selecting a method is the degree of specificity required. For example, because of the relatively low sensitivity of the *Escherichia coli* plate method for vitamin B_{12}, extensive dilution of the sample is not required. This makes it ideal for the assay of pharmaceutical products formulated using cyanocobalamin of pharmacopeial quality. This organism's low specificity, however, severely limits its usefulness in the assay of naturally occurring preparations also containing substances related to cyanocobalamin. For these, the more specific tube assays using *Lactobacillus leichmanii*, or better still *Ochromonas malhamensis*, are more suitable. The latter, however, has the disadvantage of a long incubation period, which may be as much as seven days.

Referring now to assays in general, statements may be seen in the literature that a particular turbidimetric assay is more precise than the corresponding plate assay, or vice versa. Such assertions should be treated with some reserve, as the precision of either method may be varied at will according to the design of experiment and degree of replication. It is true, however, that there may be in some cases basic limitations dependent, for example, on poor slope, as in the plate assay for neomycin described earlier.

8.2   General Considerations in Selection of Design

Publications describing newly developed assay methods for a particular substance normally give full details of procedures so that they can be followed by any analyst without need for further experimentation. The information provided includes details of dose levels of standard and sample, replication of doses, and distribution of test solutions on plates or in racks of tubes: in other words, details of assay design. The design so described is probably appropriate to the purpose for which the assay was developed. However, it may also have been influenced by restricted facilities, author's personal preference, or lack of knowledge or experience of other designs.

When the method is adopted for use in other laboratories, the purpose of assay may be different. For example, the assay originally described may have been designed for a research and development program, whereas the new purpose may be routine quality control in production. Practical details such as composition of medium and preparation of inoculum may need no modification. It is probable, however, that assay design will need to be modified. Often, unfortunately, this simple fact is not appreciated, and a particular design tends to be associated with the assay of a particular substance.

Microbiological analysts tend to be very conservative, as witnessed by the uncritical acceptance of designs such as the 5 + 1 assay described in the U.S. FDA regulations. Some analysts outside the United States of America regard this as having some sort of "official" status lacked by other designs. They overlook the fact that the FDA regulations also describe the 2 + 2 design. The International and European Pharmacopeias describe a variety of designs including the balanced 2 + 2 and 3 + 3 but exclude the un-balanced 5 + 1 design.

The analyst should always consider purpose of assay and then select an appropriate design.

From a consideration of the examples given in Chapters 2–4, the reader will already have gained some idea of the principles governing selection of an assay design appropriate to the work in hand. Now having seen in Chapters 6 and 7 the quantitative influence of design and replication on the precision of assays, it is possible to take a more detailed view of this important topic.

It must be made clear at the outset that unless the basic chemical, physical, and biological requirements for a valid assay have been satisfied, no assay design and no amount of replication can lead to a good result.

The points relevant to selection of design are:

(1) *The number of samples* with the same active ingredient to be assayed.

(2) *The nature of the samples*, the breadth of range in which it can be supposed that the potency will lie before assays. If the "supposed potency range" is narrow, the sample can be diluted so that the test solution potency is close to that of the standard test solutions.

(3) *Classification of samples*; do all samples, some samples, or none of the samples fall into a narrow "supposed potency range" category?

(4) *The required precision of the result*; is the same precision required for all samples? If many samples are required to be assayed with relatively low precision, what are the acceptable chances of an occasional result being rather wide of the true potency?

(5) *The work capacity of the team.*

8.3 Plate Assay Designs

The capability of the various designs for plate assay are revealed by con-sideration of the components of the simple formula for variance of the logarithm of the potency ratio M (Eq. (6.6)):

$$V(M) = \frac{s^2}{b^2}\left[\frac{1}{N_S} + \frac{1}{N_T} + \frac{M^2}{S_{xx}}\right]$$

(A)   The value of both s and b are dependent on the experimental conditions of the assay. That is to say, they are influenced by the nature of the test organism, its phase of growth, density of inoculum, length of prediffusion time, and temperature of incubation.

The regression coefficient b represents the calculated increase in zone diameter for a tenfold increase in dose. It follows that its numerical value is independent of the units in which dose is measured. Assay design has no influence on the value of b.

The term s^2 represents the residual variance of zone size after separation from the total variance, the known effects due to treatment and plates (petri dish assays) or treatment, rows, and columns (large plate assays using Latin or quasi-Latin squares). It is attributed to random error and is greatly dependent on the sharpness of zone boundaries and the capability of the measuring system and operator.

In contrast to b, this residual error is influenced by assay design. If the design of assay permits isolation of components of variance, for example, for rows and columns in large plate assays, or for individual plates in petri dish assays, this usually leads to a reduction in error mean squares. In Example 7, however, in which doses are applied to a large plate in an entirely random manner, separation of components for rows and columns is not possible.

With the exception of Example 8, in all the petri dish assays described here, each petri dish represents a self-contained assay and so a component for plate squares can be isolated. This reduces the sum of squares for error, which normally leads to a reduction in the error mean squares, that is, s^2. However, isolation of the component due to plates also has the effect of reducing the number of degrees of freedom associated with error squares, so that if plate to plate variation were very small, separation of this component could lead to an increase in s^2. In such a case it would be incorrect to separate the component for plates.

In Example 8, each plate is not a self-contained assay, but allowance for variation between plates is made by the inclusion of reference points from which a plate correction can be calculated. Corrected zone sizes are used in the statistical evaluation and so the plate component of variation does not appear in the calculation. The limitations of this method of correction described by Kavanagh (1974) have been mentioned in Chapter 2. Apart from this, the method is extravagant in that half of all the zones in the assay are used only for correction purposes.

Large plates were introduced with the object of making replicate observations of each treatment in an environment more uniform than can be achieved with a set of small dishes. Nevertheless, differences do remain as were described in Section 2.10. Use of randomized Latin or quasi-Latin square designs minimizes these effects. The former are naturally more efficient in balancing out unwanted influences. Responses are influenced by their position on the plate and particularly by the variation in diffusion time. These are revealed in the varying totals for rows and columns in the Latin squares. Allowance is made for this by separation of components of variation for rows and columns, which generally leads to a reduction in the value of s^2.

The inherent characteristics of the assay and the experimental technique influence $V(M)$ only through the term s^2/b^2. It follows that when all other terms are constant, the width of the confidence limits is directly proportional to s/b.

(B)   Replication of doses within a single assay influences $V(M)$ mainly through the terms N_S and N_T but also to a slight extent through the term M^2/S_{xx} as S_{xx} increases with increased replication.

As has been seen in several examples, when M is small, the term M^2/S_{xx} makes only a very small contribution compared with the other two terms within the bracket.

For any fixed total number of observations, the value of $1/N_S + 1/N_T$ is at a minimum when $N_S = N_T$. It follows then that balanced designs are inherently more efficient than unbalanced designs, that is, when the number of responses to the standard preparation is equal to that for *each* sample. Such is the case in all the $2 + 2$ assays described in Chapter 2, whether for 1, 2, 3, or 7 samples compared with one standard, and also the $3 + 3$ assay.

When the term M^2/S_{xx} can be neglected it will be seen that for balanced assays $V(M)$ is inversely proportional to the total number of observations. It follows that the standard error of M is inversely proportional to the square root of the total number of observations. Thus the precision can be increased very simply by increasing replication. The increased replication also leads to a disproportionately higher number of degrees of freedom associated with s^2, which through its effect on the value of t reduces slightly further the width of confidence limits of the potency estimate.

These combined effects are illustrated in Table 8.1, which relates width of confidence limits and replication. The table was constructed using Eqs. 6.6, 6.7, and 6.8, putting $M = 0$, a typical (constant) value for s^2/b^2 and

Table 8.1

Variation in Precision of Balanced
Parallel Line Assays According to
Replication[a]

Replication	Relative width of confidence limits
2	2.17
3	1.54
4	1.27
5	1.11
6	1.00
8	0.85
10	0.76
12	0.68
16	0.60
20	0.53

[a] The width of confidence limits
relative to that for an assay with a
replication of 6 responses per treatment is shown for both 2 + 2 assays
(for 2 samples) and 3 + 3 assays.
The replication of 6 is chosen as an
arbitrary standard as this is a commonly used replication. Figures were
calculated on the basis of petri dish
assays with 6 zones per plate.

varying values of $N_S = N_T$ and t, according to replication. A replication
of six is the arbitrary reference point for comparison. It will be seen that
increasing replication follows a law of diminishing returns for effort.

For much routine work a replication of 6 or 10 is convenient. If the total
number of plates is to be increased to about 20, it may be preferable to divide
these between two independent assays on different portions of the sample.
In this way any bias arising from heterogeneity of the sample or inaccuracy
in preparation of the test solutions may be minimized. This topic is discussed
further in Section 9.1.

It is important, however, to relate total number of zones with assay design
in any consideration of efficiency. A comparison of some balanced designs
with the FDA standard curve method first for one and then for twelve samples
is instructive. It is assumed in this exercise that petri dishes with a capacity
for six zones would be used.

In the assay of a single sample by the FDA method with the normal
degree of replication, a total of 15 plates would be used, that is, one set of

three plates for standards at each of levels 1, 2, 4, and 5 and a further set for the sample. The 90 zones would be apportioned thus:

reference zones	45
standard (nonreference) zones	36
sample zones	9

leading to

$$\left(\frac{1}{N_S} + \frac{1}{N_T}\right) = \left(\frac{1}{36} + \frac{1}{9}\right) = 0.1389$$

For effective utilization of the plates' capacity (6 zones) in a balanced assay of a single sample, a 3 + 3 design is indicated. In this case the 90 zones would be apportioned as

standard zones	45
sample zones	45

leading to

$$\left(\frac{1}{N_S} + \frac{1}{N_T}\right) = \left(\frac{1}{45} + \frac{1}{45}\right) = 0.0444$$

As width of confidence limits is approximately proportional to the square roots of these values it is clear that in this case the balanced assay will have a substantial advantage, limits being narrower by a factor of about

$$\left(\frac{0.0444}{0.1389}\right)^{1/2} = 0.57$$

An alternative way of comparison is to say that the 3 + 3 assay could have achieved about the same precision as the 5 + 1 assay using only five plates.

When a higher number of samples is to be assayed, the contrast is much less dramatic. First taking the case of the 5 + 1 assay of twelve samples, the number of plates for the standard curve remains unchanged, but an additional three plates are needed for each extra sample, making 48 plates in all. The value of $(1/N_S + 1/N_T)$ remains unchanged at 0.1389 for each sample.

Two alternative balanced arrangements come to mind for the assay of 12 samples using 48 plates:

(a) a 3 + 3 design with the 48 plates divided into 12 sets consisting of 4 plates per sample, giving for each sample

$$\left(\frac{1}{N_S} + \frac{1}{N_T}\right) = \left(\frac{1}{12} + \frac{1}{12}\right) = 0.1667$$

(b)   a 2 + 2 design with 2 samples per plate and the 48 plates divided into 6 sets of 8 plates per pair of samples, giving for each sample

$$(1/N_S + 1/N_T) = (1/16 + 1/16) = 0.1250$$

In this case where 12 samples are assayed, the 3 + 3 design seems to have nothing to commend it. The additional work in preparing extra dose levels leads to a precision lower than that of either the 2 + 2 or the 5 + 1 design. The 2 + 2 design retains its advantage over the 5 + 1 (although with a reducing margin) however much the number of samples in the test is increased.

(3)   The term M^2/S_{xx} becomes of significance only when M has a relatively large positive or negative value. This term is also dependent on replication, overall dose range, and number of dose levels.

The combined effect of M and number of dose levels is revealed by a consideration of their influence on the value of the term $[1/N_S + 1/N_T + M^2/S_{xx}]$ of Eq. (6.6). The term is now designated Z for convenience.

Nautrally such a study must refer to a constant number of observations and a constant overall dose range. It is convenient to consider a theoretical case of the comparison of two preparations by balanced designs, with a total of 120 observations and an overall log dose range of 0.6. The convenience of these parameters will be made clear by reference to Table 8.2. The overall log dose range of 0.6 corresponds closely with a practical log dose range of 0.602 for a dose ratio of 4:1. The number of observations per preparation, 60, is the lowest common multiple of 2, 3, 4, 5, and 6, the various numbers of dose levels used in this theoretical exercise.

Table 8.2

Parameters Used in Theoretical Study of the Influence of Design on the Precision of Parallel Line Assays[a]

Number of dose levels	Number of responses per treatment	Log of dose interval	Approximate dose ratio
2	30	0.60	4:1
3	20	0.30	2:1
4	15	0.20	1.6:1
5	12	0.15	1.4:1
6	10	0.12	1.3:1

[a] In this theoretical comparison of two preparations (standard and sample), 60 doses were allocated to each preparation. The overall log dose range was 0.6, corresponding to a dose range of almost exactly 4:1. The number of dose levels was varied from 2 to 6.

The lowest value of Z is obtained when sample and standard test solutions are equipotent ($M = 0$); it is the same value regardless of the number of dose levels when N_S and N_T are fixed. Putting the number of dose levels, $k = 2$, to obtain $S_{xx} = 10.8$, and putting $M = 0$, Z is calculated as

$$\left[\frac{1}{60} + \frac{1}{60} + \frac{0}{10.8} \right] = 0.033,333$$

When M is ± 0.3 (corresponding to sample test solution potencies of 50 and 200% of the standard), Z becomes

$$\left[\frac{1}{60} + \frac{1}{60} + \frac{0.09}{10.8} \right] = 0.041,667$$

Thus, the increase in M has caused Z to increase $0.041,667/0.033,333 = 1.25$ times.

The width of confidence limits is approximately proportional to $[Z]^{1/2}$, and so it follows that increasing M from 0 to ± 0.3 leads to widening the confidence limits by a factor of $[1.25]^{1/2}$, that is, by 1.118.

In a similar manner, factors have been calculated for other values of M and k within the same overall dose range. These values are illustrated graphically in Fig. 8.1. Identical results could also be obtained for any other constant number of observations. In other words, this theoretical example illustrates the general case.

The same conclusions may be reached using the general formula of Finney (1950). That is, the interval between the fiducial limits of M is proportional to

$$\left[1 + \frac{3(k - 1)}{(k + 1)} \cdot u^2 \right]^{1/2} \tag{8.1}$$

where

$$u = (M - \bar{x}_S + \bar{x}_T)/d \tag{8.2}$$

k is the number of dose levels, and d the overall log dose range.

To illustrate the influence of overall dose range on width of confidence limits, appropriate values were substituted in Eq. (8.1) to obtain the figures given in Table 8.3.

Consideration of data such as is given in Fig. 8.1 and in Table 8.3 leads to the following generalizations.

(a) Potency estimates of highest precision are obtainable when M is very close to zero, overall dose range is as high as possible, and number of dose levels is 2.

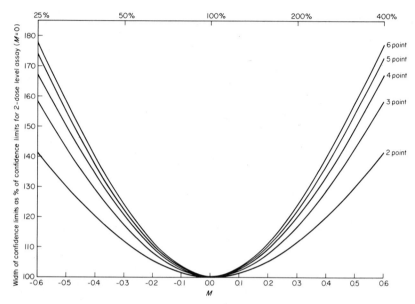

Fig. 8.1. The individual graphs for 2, 3, 4, 5, and 6 dose level balanced assays show how the confidence limits widen at increasing positive or negative values of M. Width of confidence limits is expressed as a ratio relative to the minimum value, which corresponds to $M = 0$. These curves were derived from the theoretical example with a total of 120 observations and overall log dose range of 0.6. It will be seen that with the commonly used 2 and 3 dose level assays, precision is not greatly reduced even when potencies are 0.5 or 2.0 times the standard.

Table 8.3

Influence of Overall Dose Range and M on Precision of Balanced Parallel Line Assays[a]

Log of potency ratio, M:		0.0	±0.1	±0.2	±0.3
Approximate potency ratio:		1.0	1.26 0.76	1.59 0.63	2.00 0.50
Number of dose levels	Overall dose range				
2	3:2	1.0	1.15	1.51	1.98
	2:1	1.0	1.05	1.20	1.41
	4:1	1.0	1.01	1.05	1.12
	8:1	1.0	1.01	1.03	1.05
4	3:2	1.0	1.25	1.71	2.50
	2:1	1.0	1.09	1.34	1.67
	4:1	1.0	1.02	1.09	1.20
	8:1	1.0	1.01	1.04	1.09

[a] The figures in the main body of the table represent the increasing width of confidence limits relative to that attainable when M is zero.

(b)   Precision of potency estimate is only slightly reduced at values of M of up to ± 0.1 and when overall dose range is $2:1$ or greater. For such values of M and of dose range, number of dose levels has little effect.

(c)   At values of M in the ranges $\pm(0.1$ to $0.3)$ if the overall dose range is not less than $2:1$, then results of reasonable precision are obtainable.

(d)   Even when values of M lie outside the range ± 0.3, potency estimates of acceptable precision may be obtained. However, each case should be considered on its own merits.

Consideration of this evidence alone shows that an overall dose range of at least $2:1$ should be used whenever possible and that a low number of dose levels is preferable.

8.4   The Influence of Curvature in Parallel Line Assays

In the development of any new assay procedure it is customary to determine the range in which the selected response parameter is directly proportional to the logarithm of the applied dose. A convenient starting point is a $2:1$ dose ratio, four dose level standard curve such as was shown in Example 1. This covers an $8:1$ overall range and subsequent routine work can usually be accommodated well within this range. However, many assay lines are not linear over so wide a range. Preliminary work such as this may also include variations in the level of inoculum so as to determine the most convenient conditions to achieve sharp zone boundaries and a good slope.

The basic assumption of all calculation procedures is that dose–response lines for standard and sample are both straight and parallel. This applies even in the case of those assays using only a single dose level for sample and thus providing no evidence from the data of the assay as to the nature of the sample dose–response line.

Excluding minor deviations attributable to random error, the need for response lines to be parallel is absolute. In contrast, substantial curvature (in parallel line assays) may have negligible influence on the potency estimate.

The reason for this may be seen by consideration of the nature of the curvature.

A straight line response may be expressed in the form

$$y = a + bx \tag{8.3}$$

and a curved response line in the form

$$y = a + bx + cx^2 + dx^3 + ex^4 + \cdots, \quad \text{etc.} \tag{8.4}$$

where x is log dose and y the corresponding response (zone diameter) in millimeters.

Hewitt (1977a) derived expressions in the form of Eq. (8.4) that described accurately the dose–response lines in individual agar diffusion assays over a dose range of up to about 17:1.

Using such expressions, ideal responses (freed from random variation) were calculated for hypothetical doses representing "standard" and "unknown." Substituting these "responses" in the appropriate expressions for E and F (Appendix 2) the influence of curvature (freed from random variation) on estimated potency was studied.

An expression of the form Eq. (8.4) was derived for the log dose–response line in the assay of streptomycin. It was shown that, provided that a balanced assay were used, curvature had only slight influence on calculated potency even when sample and standard potencies were very different. However, if an unbalanced design were used, bias increased markedly as "unknown" and "standard" potencies diverged. This is illustrated in Table 8.4, where "known potencies" are compared with "estimated potencies" for four different experimental designs. This streptomycin log dose–response relationship is shown in two forms in Fig. 8.2.

Table 8.4

Influence of Curvature of Response Line and Experimental Design on Estimated Potency[a]

Known[b] relative potency	2 + 2 assay design 2:1 dose ratio		2 + 2 assay design 4:1 dose ratio		3 + 3 assay design 2:1 dose ratio		5 + 1 assay design 5:4 dose ratio	
	Estimated relative potency	Bias (%)	Estimated relative potency	Bias (%)	Estimated relative potency	Bias (%)	Estimated relative potency	Bias (%)
0.500	0.500	0.00	0.492	−1.60	0.497	−0.60	0.472	−5.60
0.630	0.629	−0.16	0.622	−1.27	0.627	−0.48	0.616	−2.22
0.794	0.793	−0.13	0.788	−0.60	0.791	−0.38	0.790	−0.50
1.000	1.000	0.00	1.000	0.00	1.000	0.00	1.000	0.00
1.260	1.261	+0.08	1.266	+0.48	1.263	+0.24	1.252	−0.63
1.587	1.589	+0.13	1.600	+0.82	1.593	+0.38	1.552	−2.20
2.000	2.000	0.00	2.018	+0.90	2.005	+0.25	1.907	−4.65

[a] Exemplified by a log dose–response (zone diameter) relationship for streptomycin using *Bacillus subtilis*.

[b] This series of values for known relative potency was selected for convenience of calculation. It is a geometrical progression of ratio $1:2^{-3}$ and so includes the values 0.5, 1.0, and 2.0.

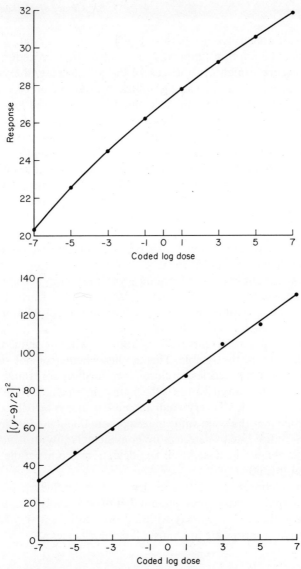

Fig. 8.2. (a) A standard response line for streptomycin in the agar diffusion assay using *Bacillus subtilis*. This is based on eight dose levels in the ratio 3:2 (log dose ratio is 0.17609). The line may be represented by the equation

$$y = 27.032 + 8.901x - 2.187x^2 + 1.108x^3 - 0.799x^4 + 0.431x^5$$

where x is log dose (relative to mean log dose which is set at zero), and y is zone diameter in millimeters.

(b) The same experimental data are plotted in the form of square of zone width (which is calculated as $[(y - 9)/2]^2$, where y is zone diameter and 9 is reservoir diameter in millimeters) against log dose. The resulting straight line is in conformity with the theory of zone formation as described in Section 2.2.

Table 8.4 demonstrates the biases at different potency ratios that arise from the false assumption (implicit in the calculation procedure) that the response lines are straight. In the case of the 5 + 1 assay, the same biases would have been found if the usual graphical procedure had been employed, using the straight line joining the calculated ideal high and low responses.

It is concluded that for the 5 + 1 assay design, bias arising from curvature is likely to be negligible only if sample potencies are not too far removed from that of the reference point. Bias would not arise if potencies were interpolated from a smooth curve drawn through the mean values of observed responses. Instead of bias we should then be faced with the old problem of drawing the best smooth curve, a problem liable to lead to greater inaccuracy than the bias that had been avoided!

In Fig. 8.2b, the same data for this streptomycin standard curve are plotted as the square of the zone width, $[(y - 9)/2]^2$ against log dose, where y is the zone diameter and 9 is the diameter of the reservoir in millimeters. The resulting straight line is in accordance with the theory of zone formation as discussed in Section 2.2. Thus, although the problems of curvature are not great, they could be avoided entirely by use of the correct relationship!

It is shown in Appendix 13 that in parallel line assays when a balanced design is used, quadratic curvature is entirely without influence on the estimated potency of the sample. This applies whether or not sample and standard test solution potencies are close. Over the short dose ranges normally used in assays, quadratic curvature is the main component of observed curvature, thus accounting for its very small influence in assays by balanced design.

It seems then that the main value in using more than two dose levels is not to detect curvature but to detect more convincingly any drift in estimated potency with change in dose level, i.e., deviations from parallelism as an indication of invalidity.

Taking into consideration this evidence of the influence of curvature together with that on dose range and number of dose levels from Section 8.3, it is concluded that a 2 + 2 assay with 2:1 dose ratio is a very satisfactory basis for design, appropriate in almost all circumstances.

8.5 Choice of Design for Plate Assay

When practical techniques including plate reading have been standardized so as to ensure accuracy and the assay design has been fixed, the attainable precision may be varied at will according to the degree of replication. Thus in theory at least, any design can be used for a high precision assay. In practice, if a reliable result is needed then it is necessary not only to control practical techniques but also to choose a design that will yield the desired precision most efficiently and therefore economically.

When high precision is not required, then again economic considerations dictate that an appropriate design be selected and superfluous replication be avoided.

The terms high precision, moderate precision, and low precision have become (perhaps particularly in Europe) part of the jargon of microbiological assay. But what is the use of a high-*precision* assay that is not also of high *accuracy*? We should be concerned not only with precision but with the overall capability of the test to give us the information we need with a *reliability* commensurate with our purpose in testing.

For present purposes requirements of an assay are designated thus:

Class 1: yielding very reliable potency estimates in the sense of high accuracy with high precision.

Class 2: economical tests yielding moderately reliable potency estimates in the sense of moderate precision and generally high accuracy.

Class 3: screening tests in which economy is of prime importance. These may include those tests used only to make the decision "accept" or "retest"; at the lowest extreme they may be little more than a qualitative test for microbiological activity.

Using this classification, designs and purpose of test may now be considered, starting with class 3 and working upward. However, dividing lines are not always distinct.

An assay design that is very appropriate for class 3 work is the large plate method, which uses only one dose level of sample and four dose levels of standard. This was illustrated in Example 7. It has an overall dose range of ratio 8:1 for the standards and so is particularly suitable when the analyst has little idea of sample potency before testing. An occasional very high or very low potency test solution does not invalidate the assay by making a false contribution to the slope when it may be far outside the range of linearity. It is useful therefore for a variety of samples in control of antibiotic extraction processes where the range of samples may include those which should display little or no activity, e.g., final fraction of a resin column eluate. The single dose level of sample does not permit any validity checks, but this is quite acceptable bearing in mind the limited aims of the test.

Other economical designs include the 2 + 2 petri dish assay for two samples and one standard or a 2 + 1 design that could accommodate up to four samples at a single dose level as well as standard at two dose levels on each plate. As a general guide, replication should never be less than four.

All these class 3 assay designs are suitable for screening large numbers of samples. They could be used for instance to check many samples of pharmaceutical products from the market with the intention of sorting out those of doubtful quality for more thorough examination later.

The 5 + 1 design illustrated by the assay of kanamycin (Example 8) can be extended very easily to any number of samples and any degree of replication. It may be used for work of class 3 or class 2. Application of test solutions to plates is a simple routine as the same pattern for distribution of two test solutions is used on every plate. However, the method has several disadvantages, some of which were mentioned in Section 2.14. The single dose level for samples does not permit any checks on validity of response. If the sample warrants a class 2 rather than a class 3 result, then it is surely worthwhile including a very simple check that the basic assumption of parallel responses is in fact correct.

The five doses of standard do theoretically permit a check for curvature, but this is not usually detectable in such a narrow overall dose range, i.e., $(5/4)^4$, which is about 2.5:1. Moreover, as has been shown in Section 8.4, curvature is unlikely to bias the potency estimate significantly. Thus a five dose standard curve is superfluous. Kavanagh (1975) advocates a simplified modification of the "FDA" method, using only two points in a 2:1 dose ratio to define a standard curve. The higher dose level serves as the reference dose but in other respects, e.g., distribution of test solutions on each plate and the use of a correction procedure, the routine is similar to the standard "FDA" procedure. This is clearly an improvement as regards both practical operations and design, since awkward dilutions and wasteful effort on the standard curve are avoided. The extra computational effort necessitated by the correction procedure remains a disadvantage, however.

It should be noted that this 2 + 1 design is not the same as the 2 + 1 design of the International and British Pharmacopeias. In the latter each plate may be considered a self-contained assay, and so it follows that a plate correction is not applicable.

For much routine work, class 2 assays using two dose levels in a 2:1 ratio are most convenient. To use more than two dose levels entails additional effort all to no good purpose. Petri dishes for two samples and a standard are appropriate for any even number of samples. A replication of about 6 or 8 is suitable. A large plate 30 cm square can be used to accommodate two samples and a standard in a 6 × 6 design or three samples and a standard in an 8 × 8 design. The latter, of course, makes more efficient use of materials and equipment. Thus if 6 samples are to be assayed, it is more efficient to use two 8 × 8 plates (128 zones) with three samples on each, than three 6 × 6 plates (108 zones) with two samples on each.

When large plates are available and many samples are to be assayed, the 8 × 8 quasi-Latin square is convenient in that it accommodates seven samples on each plate. Replication of each test solution however is only 4, that is, 8 zones for each preparation, and so this design is inherently less precise than the 8 × 8 Latin square design for three samples. In this latter design, replication is 8 for each test solution, corresponding to 16 zones for each test preparation.

In those cases when a class 1 assay is required, the qualitative nature of the material and its approximate potency are generally known in advance of the assay. Sample and standard test solutions may be prepared so as to be nearly equipotent so that M is small, with the consequence that the influence of curvature is negligible. In such circumstances there will rarely be any need to test for curvature, and so the design of choice is that which will be most efficient, normally a 2 + 2 design. However, there may be other considerations according to the number of samples to be assayed and the assay facilities. Thus, it would be uneconomical to insist on a simple 2 + 2 assay at the expense of incomplete utilization of space on the assay plate.

Some appropriate designs are as follows:

(1)   *For one sample.*   If large plates are available a 2 + 2 design using an 8 × 8 Latin square in which each test solution appears twice in each row and each column is convenient. This results in a replication of 16 for each of the four test solutions. If petri dishes having capacity for 6 zones per dish are used, then to preserve the balance and utilize the plate's full capacity a 3 + 3 design with replication of about 12 would be appropriate.

(2)   *For two samples.*   If large plates are used, then it is probably best to use them exactly as described above, making quite separate assays for the two samples. Use of the plates for a 6 × 6 design incorporating both samples in a multiple 2 + 2 assay would not utilize the space on a 30 cm square plate and would be less precise than the 8 × 8 mode of use. If petri dishes are used, then the multiple 2 + 2 mode (one set of 18 plates) or the simple 3 + 3 mode (two sets of 12 plates) should yield about the same precision, as in both cases $N_S = N_T = 36$. The former is clearly the more efficient as it avoids the extra dilutions and uses only 75% of the number of plates of the latter.

(3)   *For three samples.*   When large plates are used, it would be more efficient to use 8 × 8 Latin square designs for three samples on one plate (2 + 2 design) than to use separate plates as in (1). Precision will be determined (other factors remaining unchanged) essentially by degree of replication. Three separate assays (three plates) would result in $N_S = N_T = 32$ for each preparation. Using the multiple assay, however, the same total replication would be attained with only two plates.

Probably the most elaborate current design uses extra large plates and a 12 × 12 Latin square design. There are 12 treatments comprising two weighings each of one sample and a standard, and dilution to three dose levels. Each treatment appears once in each row and each column. The internal evidence of the assay permits checks on curvature, deviations from parallelism, and weighing or dilution discrepancies.

The main use of these designs is for the establishment of company working standards or in collaborative assays to set up new national or international standards.

Such designs are capable of yielding potency estimates with confidence limits for a single assay of within $\pm 1\%$ ($P = 0.95$). However, it has been said that this is only achieved and the rigorous statistical requirements are only met when the assay is carried out by highly skilled operators. This perhaps should be regarded as a *general* comment on the limitations imposed on assay capabilities by the skill of the operators. It has been brought to light in the case of these high-precision assays not due to any innate peculiarity of the method, but merely because its limitations have been investigated critically.

The conclusion to be drawn is that such designs are not an automatic passport to high-quality assays. Their use should not be undertaken lightly.

8.6   Tube Assays—General Considerations

Despite the fundamental difference in nature of response, tube assays for antibiotics and growth-promoting substances have points in common relevant to experimental design. The extraneous influences that may cause varying responses to replicates of the same dose have been described in Chapters 3 and 4. It is possible by meticulous experimental technique to control and minimize these unwanted influences. This is in contrast to plate assays where, even in the high precision 12 × 12 Latin square design assays, replication and randomization play an important part in compensation or balancing out unwanted influences that remain despite care taken to reduce them.

However, the contrast between these two basic methods is not absolute but is a matter of degree. While with a high standard of practical technique the required degree of replication in tube assays may be minimal, it is a fact that this standard is often not achieved and replication of three or four tubes is commonly used. Such is the case in several examples given in Chapters 3 and 4. In contrast, Kavanagh (1972) claims that with the high-precision water bath of the Autoturb® system even duplicate responses have little advantage over single responses to each test solution.

8.7   Designs for Slope Ratio Assays

For the purpose of defining principles of assay design, it is convenient to assume the ideal case in which response is directly proportional to dose. Slight deviations from linearity will not invalidate the reasoning.

The principles of design of slope ratio assays can be understood best by reference to the expression for variance of potency ratio Eq. (7.3), which may be rewritten as

$$V(R) = \frac{s^2}{b_S^2} \frac{l}{m} \tag{8.5}$$

where

$$l = S_{x_S x_S} + 2R S_{x_S x_T} + R^2 S_{x_T x_T} \tag{8.6}$$

and

$$m = S_{x_S x_S} S_{x_T x_T} - (S_{x_S x_T})^2 \tag{8.7}$$

(A)   The numerical value of b_S is dependent on the measuring system and units of measurement of response. It is also dependent on the units for measuring dose. This latter point is in contrast to parallel line assays for which the value of b is independent of the units in which dose is measured. However, it is the response to overall dose range and not b_S that influences precision. Any change in technique that leads to an increase in b_S by giving the same response for a lower dose causes the value of l/m to vary in such a way that it exactly counteracts the change in b_S, so that the value of $V(R)$ is unchanged.

(B)   The overall dose range exerts its effect on $V(R)$ through the term l/m, which varies inversely as the square of the overall dose range. It follows that a doubling of the dose range leads to halving of the range of confidence limits.

The range should be chosen so that the highest responses are as large as is consistent with precise measurement and their being in the rectilinear portion of the dose–response curve.

(C)   The term s^2 represents the residual variance of individual tube response due to random error. Like b_S, its numerical value depends on the units and method of measuring response. As these two terms occur in numerator and denominator, respectively, the net result of change in units for response measurement is nil.

(D) The terms l and m are both related to degree of replication and to number of dose levels. Increasing the total number of responses leads to a lower value of the fraction l/m and thus to higher precision. Imbalance of assay design, i.e., unequal numbers of responses for standard and sample, reduces precision relative to that which could be achieved from the same total number of responses in a comparable balanced design.

The effect of replication via t, and l/m on width of confidence limits is shown in Table 8.5 for both five point and seven point assays.

Table 8.5

Variation in Precision of Slope Ratio Assays According to Replication[a]

Replication	Relative width of confidence limits	
	Two dose levels	Three dose levels
2	1.00	1.00
3	0.71	0.74
4	0.59	0.62
5	0.51	0.55
6	0.46	0.50

[a] The width of confidence limits relative to that for an assay of the same number of dose levels and a replication of two responses per treatment is shown.

The effects of number of dose levels within a constant range and ratio of sample potency to standard potency are interdependent and so are best considered together. As in the case of parallel line assays, it is convenient to use a theoretical example in which a constant number of observations is arranged in various designs.

For a comparison of 1, 2, 3, and 4 dose level assays using a constant number of observations, 315 is the smallest possible number of observations. This is the lowest common multiple of 3, 5, 7, and 9, the number of treatments in these four different designs.

The parameters of these different assay designs are shown in Table 8.6.
Using the parameters of each design in turn and substituting a range of
values for R in Eq. (7.3), the following basic values were calculated: l/m,
which is proportional to $V(R)$; $[l/m]^{1/2}$; and $[l/m]^{1/2}/R$, which is approximately proportional to width of percent confidence limits.

The basic conclusion is very simply that maximum precision is obtained
when the number of dose levels is one (a three point assay) and R, the ratio
of estimated sample potency to standard potency, is 1.0. It is convenient
therefore to take this condition as a reference point for comparison of the
effects of variations in number of dose levels and changes in R.

Table 8.6

Parameters Used in a Theoretical Study of the Influence of Design on
Precision of Slope Ratio Assays[a]

Number of dose levels	Number of treatments	Coded doses				Replication
1	3	0	12			105
2	5	0	6	12		63
3	7	0	4	8	12	45
4	9	0	3	6	9	12	35

[a] In this theoretical comparison of two preparations (standard and
sample), 315 doses were divided equally between the treatments (including zero dose). The overall dose range was 12 coded units and the
number of dose levels was varied from 1 to 4.

The quantitative conclusions of this theoretical exercise are shown in Figs.
8.3–8.5 and in Table 8.7. In Fig. 8.3 it is seen that $V(R)/R$ increases as R
deviates from 1.0. When plotting against R on a log scale the values of l/mR
(which is proportional to $V(R)/R$) are symmetrically distributed. This form
of presentation, however, is not so useful as those in Fig. 8.4 and 8.5, which
show the changing width of confidence limits with changing values of R and
different numbers of dose levels covering the same overall range.

In Fig. 8.4, varying values of $[l/m]^{1/2}/R$ are divided by the value of this
expression for a three point assay at $R = 1$. In Fig. 8.5, the varying values of
$[l/m]^{1/2}/R$ are divided by its value for an assay with the same number of dose
levels at $R = 1$.

Both these series of graphs indicate a slight improvement in precision as
values of R exceed 1.0. However, this is misleading; it is clear that in such

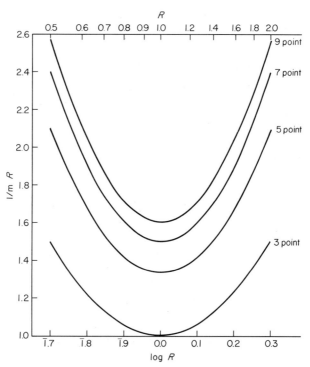

Fig. 8.3. The influence of potency ratio on precision of balanced slope ratio assays (1). The change in the ratio l/mR [which is proportional to $V(R)/R$] according to number of dose levels and value of R is shown. Each value of l/mR is expressed relative to its value for a three point assay for which $R = 1.0$.

circumstances either the dose range of the standard is less than optimum or the higher sample responses lie on a nonlinear portion of the dose response curve. Take, for example (referring to Fig. 8.5), a sample for which $R = 2.0$ ($\log R = 0.3$) and having confidence limits 87.5% of those for $R = 1.0$. If the standard test solution doses could be doubled (with responses all remaining on a linear part of the curve) and those of the sample remain unchanged, then the potency ratio would be 1.0. As explained in Section 8.7(b), a doubling of the dose range leads to halving of the width of confidence limits. Thus, a potency ratio of 1.0 and optimum dose range is by far preferable.

Wood and Finney (1946) have shown that equal distribution of tubes at all dose levels is not the most efficient. Greater precision can be obtained for a fixed number of tubes by allocating more tubes to higher dose levels. However, this complicates calculation procedures and is less convenient in practical application.

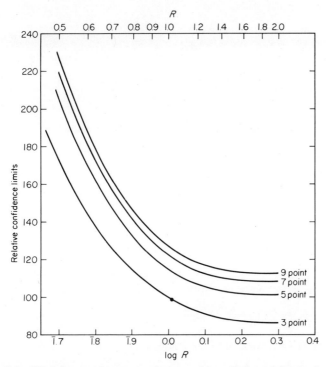

Fig. 8.4. The influence of potency ratio on precision of balanced slope ratio assays (2). It is shown how the width of percent confidence limits varies according to R and the number of dose levels. The arbitrary reference point for comparison is a three point common zero assay and a sample for which potency ratio $R = 1$.

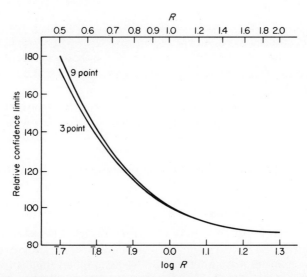

Fig. 8.5. The influence of potency ratio on precision of balanced slope ratio assays (3). The variation of relative confidence limits with R is shown for three and nine point assays. The relative confidence limits are expressed as percentage of their value at $R = 1.0$ for three and nine point assays, respectively.

Table 8.7

Influence of Number of Dose Levels on Precision of Slope Ratio
Assays[a]

	Relative widths of confidence limits	
Number of dose levels	$R = 1.0$	$R = 0.5$ or 2.0
1	1.00	1.00
2	1.15	1.18
3	1.22	1.26
4	1.27	1.31

[a] The interval between confidence limits is expressed relative
to that for a three point assay having the same number of observations, the same overall dose range, and same value of R. Figures
were calculated from the theoretical example of 315 observations.

8.8   Influence of Curvature in Slope Ratio Assays

The slope ratio calculation assumes that response lines are straight. In
contrast to parallel line assays, even slight curvature may cause a significant
bias, with estimated values of R closer to 1.0 than the true value of the
potency ratio.

The extent of the influence of curvature was demonstrated by Hewitt
(1977b) using procedures analogous to those described for parallel line assays
in Section 8.3. It is illustrated here by reference to the assays of nicotinic acid
(slight curvature) and thiamine (strong curvature) using the data of Önal
(1971) Fig. 3.2.

Expressions of the form of Eq. (8.4) were derived from responses to zero
control and the lower standard dose levels only, thus avoiding the more
strongly curved regions of the response lines. Using these expressions, ideal
responses (i.e., responses lying exactly on the curve) were calculated for
"unknown" preparations having potency ratios relative to the standard of
0.5, 0.8, 1.25, and 1.5.

Curves for each "unknown" preparation were drawn plotting calculated
response against coded dose (Fig. 8.6a,b). The "responses" were substituted
in turn in the appropriate form of the Bliss Eq. (3.17 and 3.18) for one unknown preparation at two dose levels to calculate apparent potency ratios.

The discrepancies arising from the false assumption (implicit in the calculation procedure) that the dose–response lines are straight are shown in
Table 8.8. It is clear that the slope ratio method of calculation should be

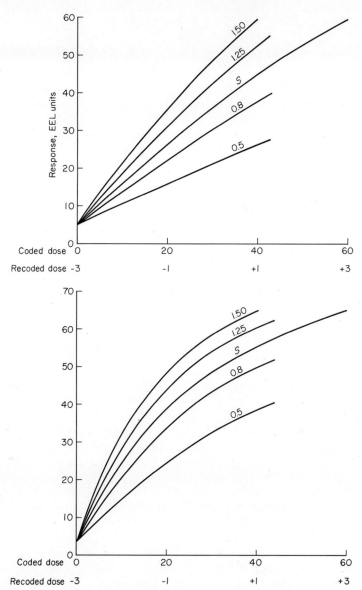

Fig. 8.6. Dose response curves for (a) nicotinic acid and (b) thiamine. The curves marked S are based on the same experimental data as shown in Fig. 3.2 for the range of coded dose 0 to 60. Expressions representing these data were derived as:

for nicotinic acid

$$y = 35.963 + 9.271x - 0.413x^2 - 0.021x^3$$

and for thiamine

$$y = 48.300 + 7.865x - 1.377x^2 + 0.288x^3 - 0.020x^4 - 0.003x^5$$

where x is the dose recoded so that 0 becomes -3 and 60 becomes $+3$, and y is the calculated response. Using these expressions, points were calculated to represent responses of preparations having potencies 0.5, 0.8, 1.25, and 1.50 of that of the standard. These calculated responses are plotted at the same nominal potencies as the standard, thus giving a series of dose–response curves.

Table 8.8

Discrepancies Arising from Use of the Slope Ratio Calculation When
Response Lines are Curved[a]

True potency ratio	Nicotinic acid		Thiamine	
	Calculated ratio	Bias %	Calculated ratio	Bias %
0.50	0.524	+4.7	0.632	+26.4
0.80	0.817	+2.2	0.873	+9.1
1.00	1.000	0.0	1.000	0.0
1.25	1.212	−3.0	1.134	−9.2
1.50	1.348	−9.4	1.251	−16.6

[a] The corresponding response lines are shown in Figs. 8.6a and
8.6b.

applied only if response lines are straight, or if curvature is very slight and
sample responses lie very close to those of the standard. If these conditions
are not met, then a purely graphical method should be used. In fact, in the
original work of Önal, sample potencies were obtained by interpolation from
the standard curves

8.9 Choice of Design for Slope Ratio Assays

The selection of a practicable assay design should be based on a considera-
tion of the theoretical aspects and the influence of curvature described in
Sections 8.7 and 8.8, respectively, and the general factors enumerated in
Section 8.6.

For a fixed total number of responses over a fixed range of doses, an
increase in the number of dose levels (and necessarily a decrease in the
replication at each level) reduces precision. If efficiency of an assay is reckoned
in terms of precision attainable with a given number of observations, then
the most efficient design would include observations at only three points.
These are one dose each for standard, unknown, and zero control. Although
efficient, such a design does not permit any validity checks. A five point
design in which doses are 0, x_S, $2x_S$, x_T, and $2x_T$ (zero control, standard, and
unknown preparations, respectively) permits a check whether the two response lines intersect at zero dose. As explained in Section 3.5, even though
lines intersect at a point corresponding to zero dose, this point of intersection
does not necessarily coincide with the observed response to the zero dose
control.

Failure of the lines to intersect at zero dose would indicate either curvature or an invalid assay. It would be necessary to repeat the test with more dose levels so as to acertain whether the comparison of the preparations were basically invalid or whether a valid potency estimate could be made graphically from curved response lines.

In the routine quality control of pharmaceuticals where supposed potency can be reasonably expected to be not far removed from actual potency, then the 5 point common zero assay or the corresponding multiple two dose level assays seem ideal in that they provide for a simple check on validity with minimum effort.

When natural products are being examined and there is a likelihood of potencies being far removed from standard then the series of doses suggested by Bliss (1952) seems very convenient: for a four dose level assay, four levels of standard solution 1.0, 2.0, 3.0, and 4.0 per tube and six levels of sample 0.5, 1.0, 1.5, 2.0, 3.0, and 4.0, thus permitting an appropriate selection of four dose levels, either 0.5, 1.0, 1.5, and 2.0 or 1.0, 2.0, 3.0, and 4.0. Alternatively, eight levels of sample may be set up at 0.2, 0.4, 0.6, 0.8, 1.2, 1.6, 2.4, and 3.2 from which three possible selections may be made:

either	0.2,	0.4,	0.6	and	0.8
or		0.4,	0.8,	1.2	and	1.6
or		0.8,	1.6,	2.4	and	3.2

8.10   Tube Assays for Antibiotics

The popularity of graphical techniques for antibiotic tube assays has already been mentioned. Interpolation at one sample dose level from a standard curve has the advantages of simplicity and absence of any restrictions imposed by assay design. In view of the day to day variability of median responses and the possible need to discard some observations, there seems little doubt that this procedure will continue to be widely used. It has the inherent disadvantages, however, of inefficiency and absence of validity checks. Despite these disadvantages, it is a convenient procedure for routine work such as screening large numbers of samples by a public laboratory in the search for low potency pharmaceutical products. Because of the many alternative ways of computation it is not so easy to define ideals in design as was the case with agar diffusion assays and tube assays for growth-promoting substances. However, when a reliable estimate of potency is wanted then it is desirable to use a computation method that avoids the use of graphs and perhaps also permits statistical evaluation.

Computation methods may be divided into two main groups: (1) those in which some function of the response plots linearly against dose, and (2) those in which some function of the response plots linearly against logarithm of dose.

The first group includes Examples 13 and 15 where the relationship log absorbance versus dose is used (Figs. 4.4c and 4.6b, respectively) and Example 14, where either probit or angular transformation of response versus dose is used (Figs. 4.5e and 4.5f). These may be calculated as slope ratio assays. Clearly, for ease of calculation and efficiency of design, doses should be evenly spaced on an arithmetic scale and not as a geometrical progression as in Example 15, an assay that was not designed to be treated as a slope ratio assay.

The criteria for choice of assay design when a slope ratio relationship is to be used are then generally similar to those described in Section 8.7 with respect to assays for growth-promoting substances. In contrast however, a larger number of dose levels would be needed to permit a check on curvature and to use the widest possible range of linear response. Perhaps four to six dose levels would be convenient.

The second dose response type group is illustrated by Example 13, using probit transformation of response versus log dose (Fig. 4.4d) and Example 15, using probit or angular transformation of responses (Figs. 4.6c and 4.6d, respectively). In these cases the criteria for choice of design are generally as for parallel line assays by the agar diffussion method but it is necessary to use more dose levels at closer intervals, so as to ensure that a portion of the response curve with a suitable slope is covered adequately.

References

Bliss, C. I. (1952). "The Statistics of Bioassay." Academic Press, New York.
Finney, D. J. (1950). *In* "Biological Standardization" (J. H. Burn, ed.). Oxford Medical Publications, Oxford.
Garrett, E. R. *et. al.* (1971). "Progress in Drug Research," Vol. 15. Birkhauser Verlag, Basel and Stuttgart.
Garth, M. A. (1972). Personal communication.
Hewitt, W. (1977a). In preparation.
Hewitt, W. (1977b). Unpublished observations.
Kavanagh, F. W. (1974). *J. Pharm. Sci.* **63**, 1459.
Kavanagh, F. W. (1975). *J. Pharm. Sci.* **64**, 1224.
Önal, Ü, (1971). Personal communication.
Simmons, R. J. (1972). *In* "Analytical Microbiology" (F. W. Kavanagh ed.), Vol. II. Academic Press, New York.
Wood, E. C., and Finney, D. J. (1946). *Q. J. Pharm. Pharmacol.* **19**, 112.

CHAPTER 9

REPEATED ASSAYS, SPECIFICATIONS, AND REPORTS

9.1 Replication of Assays

Notwithstanding the evidence that has been given in earlier chapters, it remains a widely accepted concept that microbiological assay is innately liable to give variable results. To counter this, duplicate determinations seem to be a standard routine in most laboratories. The duplicate results are compared and an arbitrary decision is made as to whether they are in sufficiently close agreement. If the discrepancy appears too great, then the assay may be repeated again in duplicate. In contrast, the same laboratory may accept the results of a single chemical assay without question, provided that it lies within the required limits or is about the expected level.

While it is not suggested here that this approach to interpreting results is entirely unreasonable, a less empirical approach may often lead to a substantial saving of effort and also speed up the output of assay reports.

It is instructive to consider what is achieved by repeated assays. Is the increased precision necessary? What are the real sources of variation that are supposedly minimized by taking the mean of two or more assays? Is there any advantage in repeating assays on different days as distinct from doing replicate assays on one day?

Considering first replicate assays done on the same day, sometimes these may not be truly independent tests but merely the application of the same test solutions to other sets of plates or tubes. In effect then, two or more sets such as these form a single assay with increased replication and therefore with improved precision as described in Chapter 8.

Naturally any bias attributable in any way to the test solutions is uninfluenced by the degree of replication. Thus, increased precision is not accompanied by increased accuracy.

When the replicate tests are based on separately prepared test solutions, then increased precision may be accompanied by increased accuracy. In routine work it would only be the sample test solution that would be separately prepared, as it is customary to use a single master standard solution for all tests over a period of from one or two days to a week or more, according to the stability of the solution.

If reasonable weights and volumes are used such as are described in Appendix 6, and due allowance made when necessary for moisture content of the standard, then there is no good reason why the standard solution should not be accurate to within 0.1 or 0.2%. The use of the same standard solution should not therefore introduce any bias.

However, errors can occur even in this simple step if technicians are inadequately trained and supervised. Simpson (1963) in his contribution to "Analytical Microbiology," Volume I, places great emphasis on the training of staff.

If the sample is an uncompounded form of the "pure" active substance, then this too should be prepared with an accuracy equal to that of the standard. In such cases then, the replicate preparation of sample solutions serves mainly as a check against the possibility of gross errors, errors that should not occur in a well-regulated laboratory. Certainly gross errors in such simple operations should be no more likely a feature of microbiological assay than of chemical assay. If they are more likely in microbiological assay, then this is an indication of inadequate training and supervision. It probably also indicates a fundamental failure to accept microbiological assay as a branch of pharmaceutical analysis that requires all the skill and attention to detail as is taken for granted in the conduct of chemical assays.

Thus in the examination of a sample of the "pure" active substance, while an independent replicate assay will increase the *precision* of the potency estimate, there should be no room for improvement on the *accuracy* of a single assay!

When the sample consists of the active substance in a naturally occurring mixture or in a compounded pharmaceutical form, the situation may be very different. Heterogeneity of the material may be a major cause of differences between replicate assays. If so, is the purpose of the assay merely to obtain an average figure or is it to determine whether different samples differ significantly in potency? Clearly the latter calls for much more work.

If sample preparation involves an extraction process, then this might be an important source of error. Independently prepared test solutions, possibly prepared by different operators, might be a noticeable source of variation. The supervisor must then use his knowledge first of the precision of the assay (excluding extraction variations), and then the difficulties of the extraction procedure as well as the skill and experience of the operators to answer the questions:

Are the differences to be attributed to the random variation of the assay? If so, the results may be averaged. If not, does the lower result indicate incomplete extraction? If so, then one or two repeat assays are indicated.

These are the normal problems of analytical chemistry and are not biological in origin!

Finally, the question of assays repeated on a different day must be considered. Certainly the response of the assay system to a particular antibiotic or g.p.s. may vary from day to day. As mentioned in Chapter 8 this is a more serious problem in tube than in plate assays.

However, provided that the following conditions are met, then sample and standard should be influenced in exactly the same manner, and so day to day variation in estimated potency should be no greater than in independent assays carried out on the same day. The conditions are:

(1)   The experimental design is such that standard and sample test solutions are handled so far as possible in an identical manner and randomization is used to minimize any bias that might arise from inevitable differences in position of the test solution in plate or rack and time of application.

(2)   Sample and standard test solutions are qualitatively identical and differ only in concentration.

The first of these conditions is only a statement of routine good practice and so there is no reason why it should not be achieved.

The second, however, may not be so simple or may even be unattainable.

If the sample is either a crude form or a compounded form of the active principle, then failure to remove interfering substances by an adequate purification process could lead to varying responses from day to day. As none of the individual results would be reliable, then it would be necessary to devise an improved extraction procedure.

However, when the active principle is a mixture of two or more related substances, then day to day differences in response may be an inevitable feature of the assay if sample and standard contain different proportions of these substances.

Neomycin is a typical example of this problem. The best that can be done is to assay samples on several days and report the mean.

Fortunately many antibiotics, although mixtures of related components consist predominantly of only one constituent, and so day to day variation of assay conditions does not lead to any perceptible influence on estimated potency.

9.2 Collaborative Assays

The value of collaborative assays is not only that a more precise comparison is achieved, but also that differences in methods or operations between laboratories may bring to light discrepancies that could not be observed in a single laboratory. Investigation of such discrepancies may reveal fundamental differences in response to the two preparations under differing assay conditions.*

For this reason, in a collaborative assay for the establishment of an antibiotic reference standard, it is desirable to use more than one test method, e.g., different test organisms and preferably both diffusion and tube assays. In this way any qualitative differences between the two preparations would be revealed and might indicate that the material proposed as the new reference preparation was in fact unsuitable for the purpose.

For success in a collaborative assay, adequate preliminary consultations between participants and careful planning are of utmost importance.

The following routine has been found to smooth out many (but not all) of the difficulties in collaborative assays on a national scale:

(1)   Consultation between participants and the organizer to discover details of method that are in routine operation in each laboratory, e.g., plate or tube, test organism, media, pH, incubation temperature and duration, method of killing organism (tube assays), method and precision of measuring responses, and (if applicable) method of moisture determination.

(2)   Agreement between participants on which methods will be used so as to ensure that the whole collaborative exercise will include at least two different methods. Individual laboratories, however, may elect to use only one method.

(3)   Agreement between participants and the organizer as to what designs are acceptable for purposes of evaluation.

(4)   Issue of detailed written guidance by the organizer taking into consideration the principles agreed upon and points that appear to need stressing. Experience shows that this guidance should include notes on weights and dilutions (such as given in Appendix 6), on precision of measurement of responses, and on reporting experimental data. Weighings need to be reported fully as gross, tare, and net, because mistakes are sometimes made in this first simple arithmetical step!

The value of a preliminary trial prior to the real exercise cannot be overemphasized. This brings to light the various interpretations of the written guidance (which had been thought to be unambiguous) or perhaps shows up points that have been overlooked. At this stage it is possible to make tactful suggestions to participants for modification of techniques or reporting, thus possibly avoiding the embarrassment of having to exclude a laboratory's contribution in the final test.

---

* Alternatively basic faults in technique in one or more laboratories may be discovered. This is a useful secondary benefit of collaborative tests.

9.3 Combination of Replicate Potency Estimates

When a sample has been assayed on two or more occasions the simplest calculation of the mean is to obtain the sum of the estimates and divide by the number of estimates. This method gives the simple mean. While this simple mean may not coincide with the best estimate of the mean, it is often applicable with negligible error in the routine work of a single laboratory. This will be discussed in more detail in Section 9.4.

For nonroutine work, however, such as the establishment of a laboratory working standard, or when several laboratories participate in a collaborative assay as, for example, in the establishment of a national or international reference standard, then more elaborate procedures are needed. Calculation of a weighted mean takes into consideration the varying precision of the individual assays and gives more weight to those of higher precision. The method of Perry (1950) is used here.

The weighted mean potency is calculated via the equation

$$\bar{M} = \sum WM / \sum W \tag{9.1}$$

where the individual values of W, the weights attached to each individual value of the potency estimate, are $1/V(M)$, and $\bar{M}$ is the logarithm of the weighted mean estimate of potency.

The variance of $\bar{M}$ is given by

$$V(\bar{M}) = 1/(\sum W) \tag{9.2}$$

and the standard error by

$$s_{\bar{M}} = [V(\bar{M})]^{1/2} = [\sum W]^{-1/2} \tag{9.3}$$

Approximate limits of error of M are then obtained by

$$\bar{M} \pm ts_{\bar{M}} \tag{9.4}$$

This method is stated to be satisfactory provided that the individual potency estimates are homogenous. That is to say, the distribution of the estimates is consistent with their individual variances.

A test for homogeneity of the potency estimates was introduced by Humphrey *et al.* (1952) using the approximate relationship

$$\chi^2 = \sum W(M - \bar{M})^2 \tag{9.5}$$

The value of χ^2 obtained by substituting in Eq. (9.5) is compared with limiting values of χ^2 tabulated according to degrees of freedom and probability. Tables of χ^2 are given by Fisher and Yates (1963). An abridged version is given in Appendix 11.

The number of degrees of freedom is one less than the number of potency estimates. If the value calculated for χ^2 indicates a value of P (probability) less than 0.05, this is taken as evidence of heterogeneity. That is, the spread of results is greater than would be expected from the internal evidence of each assay. In such circumstances, the confidence limits calculated by Eq. (9.4) would be an overoptimistic estimate. To counteract this, Humphrey introduced a heterogeneity factor H defined as

$$H = \chi^2/\text{d.f.} \tag{9.6}$$

The total weight of the assays in the group $\sum W$ is divided by H to give a corrected weight, which is then used in the calculation of confidence limits of $\bar{M}$.

It is important to note that these calculation methods assume that samples have been diluted to give test solutions of nominal potency *identical* with the actual potency of the standard test solution. If this is not the case, then a "corrected" value of M can be calculated back from the potency estimate. This is made very clear by taking an extreme case, Sample 1 in the neomycin assay of Chapter 2, Example 5.

The standard high dose test solution has an actual potency of 60.4 U/ml. The nominal potency of Sample 1 high dose test solution is given by

$$(216.4 \times 646 \times 10)/(200 \times 100) = 69.9 \quad \text{U/ml}$$

In the routine calculation of potency because of the higher concentration of the sample test solution, M is a positive value $+0.03927$, despite Sample 1 being less potent than the standard.

If the concentrations as micrograms of neomycin sulfate per milliliter had been the same for both solutions, then the value of M would have been negative. This negative value is easily calculated from the potency of the standard and estimated potency of the sample as

$$\log 646 - \log 607.3 = 2.8102 - 2.7834 = -0.0268$$

This "corrected" value, -0.0268, is the one that must be used in the heterogeneity test and in determining the weighted mean.

The calculations to test for heterogeneity and to obtain the heterogeneity factor, the weighted mean, and its confidence limits may now be illustrated by means of Example 19, in which estimates from six assays of a penicillin sample are combined.

Example 19: Calculation of the Best Estimate of Mean Potency and Confidence Limits from Six Individual Assays of a Sample of Potassium Benzylpenicillin Using the Agar Diffusion Method and *Sarcina lutea* (ATCC 9341) as Test Organism

The basic data for the calculation are given in Table 9.1. This is essentially the values of M (corrected for differences in concentration of the test solutions) and the variance of each value of M. Calculation of $\sum W$ and $\sum WM$ is shown in Table 9.2. This leads by Eq. (9.1) to

$$\bar{M} = +895.22/97{,}431 = +0.0092$$

Nominal potency of the sample was 1595 IU/mg so that weighted mean potency is estimated as

$$\text{antilog}[(\log 1595) + 0.0092] = \text{antilog}[2.2028 + 0.0092] = 1629 \quad \text{IU/mg}$$

Table 9.1

Basic Data for Calculation of the Best Estimate of Mean Potency and Its Confidence Limits from Six Individual Estimates (Example 19)[a]

Assay no.	Potency IU/mg	Confidence limits IU/mg ($P = 0.95$)	M "corrected"	$V(M)$
1	1629	1570–1691	+0.0092	0.0000628
2	1648	1599–1699	+0.0143	0.0000409
3	1613	1541–1688	+0.0048	0.0000930
4	1586	1530–1644	−0.0025	0.0000578
5	1553	1480–1631	−0.0115	0.0001050
6	1693	1637–1753	+0.0261	0.0000520

[a] Total degrees of freedom for the six assays is 150.

Table 9.2

Calculation of the Weighted Mean Logarithm of Potency Ratio from Six Individual Estimates (Example 19)

Assay no.	$V(M)$	$W = 1/V(M)$	M	WM
1	0.0000628	15,835	+0.0092	145.6820
2	0.0000409	24,468	+0.0143	349.8924
3	0.0000930	10,749	+0.0048	51.5952
4	0.0000578	17,625	−0.0025	−44.0625
5	0.0001050	9,524	−0.0115	−109.5260
6	0.0000520	19,220	+0.0261	501.6420
		$\sum W = 97{,}431$		$\sum WM = +895.2231$

$$\bar{M} = \sum WM / \sum W = +895.2231/97{,}431 = +0.0092.$$

Confidence limits are then obtained via the following steps:

$$V(\bar{M}) = 1/97{,}431 = 0.00001027 \quad \text{(by Eq. (9.2))}$$
$$s_{\bar{M}} = (0.00001027)^{1/2} = 0.003203 \quad \text{(by Eq. (9.3))}$$

Log confidence limits are given by Eq. (9.4) as

$$+0.0092 \pm 1.98 \times 0.003203 = +0.0029 \quad \text{to} \quad +0.0155$$

($t = 1.98$ for d.f. 150 and $P = 0.95$).

The corresponding confidence limits as percent of the nominal potency (1595 IU/mg) are then 100.7–103.6%, which expressed as potency may be reported as "potency is estimated as 1629 IU/mg, limits 1606 to 1653 IU/mg ($P = 0.95$)."

Table 9.3

Test for Heterogeneity of Six Individual Potency Estimates of a Single Preparation by the Modified χ^2 Test of Humphrey (Example 19)[a]

Assay no.	M	$(M - \bar{M})$	$(M - \bar{M})^2 \times 10^8$	W	$\chi^2 = W(M - \bar{M})^2$
1	+0.0092	+0.0000	0	15,835	0.000,000
2	+0.0143	+0.0051	2,601	24,468	0.641,094
3	+0.0048	−0.0044	1,936	10,749	0.208,101
4	−0.0025	−0.0117	13,689	17,625	2.416,382
5	−0.0115	−0.0207	42,849	9,524	4.080,939
6	+0.0261	+0.0169	28,561	19,220	5.489,424
				Total:	12.835,940

[a] The values of $(M - \bar{M})$ refer to deviations from the weighted mean and not the simple mean. It follows that $\sum(M - \bar{M})$ is not necessarily equal to zero. In column 4, $(M - \bar{M})^2$ is multiplied by 10^8 for convenience of presentation only.

Table 9.3 shows the calculation of χ^2 in the test for heterogeneity. There are six potency estimates and therefore $6 - 1 = 5$ degrees of freedom. The values given in tables of χ^2 (Appendix 11) for 5 d.f. are:

$$\text{when} \quad \chi^2 = 11.07, \quad P = 0.05$$
$$\text{when} \quad \chi^2 = 15.09, \quad P = 0.01$$

The value of χ^2 found, 12.836, corresponding to a value of P less than 0.05 is taken as evidence of heterogeneity.

As the estimates are shown to be heterogeneous, it follows that the confidence limits of the mean estimate as initially calculated (1606–1653) are too narrow. Humphrey's correction factor H is applied to calculate a corrected weight and new confidence limits. Using Eq. (9.6),

$$H = 12.836/5 = 2.567$$

Corrected weight is given by

$$\sum W_C = [\sum W]/H \tag{9.7}$$
$$= 97,421/2.567 = 38,000$$

Corrected confidence limits are then obtained via Eqs. (9.2), (9.3), and (9.4) using the corrected weight:

$$0.0092 \pm 1.98[38,000]^{-1/2} = 0.0092 \pm 1.98[0.00513]$$
$$= \bar{1}.9987 \quad \text{to} \quad 0.0197$$

These values correspond to limits of 0.997–1.047 of the nominal potency, or 1629 IU/mg, limits 1591–1669 IU/mg ($P = 0.95$).

This group of six results represents the contribution of one laboratory using one test organism in a collaborative assay for the establishment of a national reference standard (Turkey, 1969). Using the same mathematical procedures, the weighted means of other groups of tests and, when appropriate, their corrected weights were calculated. These were then combined to give a grand weighted mean.

Thus the results of 32 comparisons of the proposed National Standard Reference Material with the International Biological Standard, carried out by four laboratories using two test organisms, were combined to give a grand weighted mean of 1595 IU/mg. Confidence limits were 1587–1604 IU/mg ($P = 0.95$).

9.4 Combination of Replicate Potency Estimates—Simplified Methods

In the routine work of many laboratories, it is neither normal practice nor is it necessary to calculate confidence limits from the internal evidence of each assay. Provided that conditions of test, method of measurement of observations, and personnel do not change, then precision should not change appreciably from test to test and from week to week.

If the precision generally attainable by a particular method has been demonstrated by evaluation of previous assays, this information can be used as a basis for routine estimation of approximate confidence limits.

Lees and Tootill (1955) introduced a simple check on the validity of an assumed standard error, which may be used when two or more independent assays have been carried out. This test is known as the "range/mean" test. The value of the fraction

$$\frac{\text{(highest potency estimate)} - \text{(lowest potency estimate)}}{\text{(simple mean of all potency estimates)}}$$

is obtained. This calculated value is then compared with the corresponding tabulated value, which relates to the assumed percent standard error and the number of individual assays.

If the tabulated value is not exceeded, then the simple mean potency estimate may be reported. If the tabulated value is exceeded, then the spread of results appears to be too great to be consistent with the expected precision of this assay. This may indicate gross errors or general deterioration of working routines. The reason for the apparent discrepancy should be sought by reexamining the entire operation. Possible nonhomogeneity of the original sample should be taken into account.

If necessary the assay should be repeated.

If the simple mean of the two or more estimates is acceptable on the basis of the range/mean test, then its percent standard error is obtained by

$$s_m = s_a/(n)^{1/2} \qquad\qquad (9.8)$$

where s_a is the assumed percent standard error of a single assay, s_m the approximate percent standard error of the simple mean of all the assays, and n the number of assays. Approximate percent confidence limits are reported as

$$\pm 2 \times s_m \qquad (P = 0.95)$$

In Example 20 the method is illustrated by application to the same set of six potency estimates on a sample of potassium penicillin as were used in Example 19.

An extended version of the range/mean table of Lees and Tootill is given in Appendix 12.

Example 20: Simplified Calculation of Mean Potency and Confidence Limits from Six Individual Estimates of the Potency of a Sample of Potassium Penicillin

The individual potency estimates are listed in Table 9.1. From this list the following are obtained:

$$\text{range of potency estimates} = 1693 - 1553 = 140 \quad \text{IU/mg}$$
$$\text{simple mean of all estimates} = 1620 \quad \text{IU/mg}$$
$$\text{range/mean} = 140/1620 = 0.087$$

The expected standard error is 3%. From the tabulated values in Appendix 12, the limiting value for six tests and a standard error of 3% is 0.121. As the value found (0.087) is less than the corresponding tabulated limiting value (0.121) the assumption of a 3% standard error is justified.

The standard error of the mean (as a percentage) is then obtained by Eq. (9.8) as

$$3/(6)^{1/2} = 3/2.45 = 1.23\%$$

and 95% confidence limits deviate by approximately plus or minus twice this value from the simple mean; i.e., approximate confidence limits are 1620, ± 40 IU/mg ($P = 0.95$).

9.5   Specifications for Antibiotics

Materials are assayed for one of two basic reasons: (1) to provide technical or medical information, or (2) to make a decision as to whether the sample conforms to a defined standard.

While the acceptable degree of precision varies according to circumstances and purpose of the test, it is almost always necessary to have some idea of the precision of the potency estimate. In case (2) however, the report usually has some legal significance whether the test is carried out in the laboratory of a producer or wholesaler or whether in the laboratory of a regulatory agency such as the Government Analyst's Department.

For many years, perhaps the most widely accepted standards for antibiotics in international commerce have been those of the British (B.P.), International (Ph. Int.), and United States Pharmacopeias (U.S.P.). The latter refers to the U.S. Code of Federal Regulations. The 1973 edition of the British Pharmacopeia accepts the antibiotic standards of the European Pharmacopeia (Ph. Eur.). In fact, the latter has to a very large extent adopted the standards and form of expression of the B.P. 1968. The B.P., Ph. Eur., U.S.P., and Ph. Int. differ in standard and method of expression.

Although superficially these standards may appear very different, closer examination reveals that they are in effect quite similar. For example, the apparently low standard of the U.S.P. for streptomycin sulfate (minimum potency 650 µg/mg) refers to the *estimated potency* on the material "as is," whereas the apparently much higher standard of the B.P. (720 IU/mg) refers to the *upper fiducial limit of the estimated potency* calculated on the basis of the dry material.

Pharmacopeial monographs define the minimum acceptable standards for the product as it reaches the consumer. Thus they are the standards accepted by the government analyst in his role of guardian of official standards.

In contrast, these monographs are not intended as manufacturing specifications and are often unsuitable for direct adaptation by the industrial analyst. This is a general statement applicable to tests by chemical, physical, and biological methods. However, it is stressed by the B.P. 1973 in Appendix XVA in the case of microbiological assays by the note that is reproduced here with the permission of the Controller of her Britannic Majesty's Stationery Office:

> Guidance to Manufacturers.   The required minimum precision for an acceptable assay of any particular antibiotic or preparation is defined in the appropriate monograph in

the paragraph on the assay. This degree of precision is the minimum acceptable for determining that the final product complies with the official requirements and may be inadequate for those deciding, for example, the potency which should be stated on the label or used as the basis for calculating the quantity of an antibiotic to be incorporated in a preparation. In such circumstances, assays of greater precision may be desirable, with for instance fiducial limits of error of the order of 98 to 102 percent. With this degree of precision, the lower fiducial limit lies close to the estimated potency. By using this limit, instead of the estimated potency to assign a potency to the antibiotic either for labelling or for calculating the quantity to be included in a preparation, there is less likelihood of the final preparation subsequently failing to comply with the official requirements for potency.

The rationale of the form of expression of standards introduced in the 1966 Addendum to B.P. 1963 was explained by Lightbown (1973) in an address to the European Pharmaceutical Inspection Convention Seminar in Edinburgh. Manufacture of antibiotics in Britain is strictly controlled under the Therapeutic Substances Regulations (T.S.R.).* Quoting from Lightbown's address:

[The] obligation on the manufacturer to ensure that the material he issues complies with the minimum specification, in that the True Potency is above the minimum, was spelt out in the revision of the Therapeutic Substances Regulations in 1966 which required that the lower fiducial limit of the manufacturer's assay should be above the stated minimum potency before the material could be legally issued. Simultaneously the BP assays for the same antibiotics that were controlled under TSR were restructured (Add 1966) so that only material with a True Potency less than the same minimum value specified in the Therapeutic Substances Regulation would be rejected [by the regulatory authority] as not complying with the requirements of the pharmacopoeia. This was ensured by the requirement that the upper fiducial limit of the pharmacopoeial assay should be less than the minimum potency before material is rejected—in which case the True Potency is ($P = 0.95$) less than the minimum potency. A pharmacopoeial assay must have a minimum precision of $\pm 5\%$ to be acceptable but it may be more precise than this.

Also quoting from this address:

The complementary procedures of manufacturer working to Therapeutic Substances Requirements whilst the [Regulatory Authority] Control Laboratory checked on the basis of the BP specification has worked satisfactorily for the last 7 years in the U.K. in relation to material manufactured at home and abroad.

9.6   Official Standards—a Practical Approach

Not infrequently, facilities and personnel in the laboratories of regulatory authorities are insufficient to permit a full examination of all the samples that are (or should be) submitted for analysis.

The country's interests will not be best served by the precise determination of potency of a small number of samples, the majority of which are in any case likely to be found satisfactory. More benefit will accrue from detection

* The TSR requirements have been superceded recently by implementation of the Medicines Act, 1968.

of substandard products by the screening of large numbers of samples from many manufacturers, which have been sampled from a variety of places and thus subjected to varying storage conditions.

A low precision assay may permit this to be achieved. Perhaps 80 or 90% of the samples examined would be accepted as apparently conforming to standard on the basis of a single assay. While a few slightly substandard samples may pass undetected, in a well-designed screening program there would be little chance of a seriously deficient sample escaping detection. Those samples that failed to pass this initial screening would then be re-assayed on one or more occasions, perhaps using a design of higher precision, so that the final verdict could be made with sufficient confidence.

In screening methods, work may be reduced to the minimum consistent with the objectives. For example, in assaying capsules of tetracycline, rather than follow exactly the directions of the B.P. and determine the weight of contents of the individual capsules, it would be adequate to weigh indivi-dually, say, five whole capsules and then transfer these to a blender for pre-paration of the test solution. This procedure makes the reasonable assumption that weight variation is attributable mainly to the content of the capsule and not its shell. In the examination of injectable products, however, it is neces-sary to weigh or measure the total content for the reasons explained in Section 1.7.

Illustrating these principles by means of streptomycin sulfate for injection, the potency of this substance is defined by the Ph. Eur. in terms of the material dried to constant weight. It follows that in countries where the Ph. Eur. has legal status, moisture content must be taken into account if an adverse report is to be issued. For initial screening purposes, however, the moisture determination may be omitted and allowance made for a "probable moisture content" of say 3%. If this probable moisture content is greatly exceeded, then it will lead to a low estimated potency, perhaps sufficiently low to justify repeat assays and determination of actual moisture content.

It is very unlikely that moisture content would be sufficiently below 3% to invalidate this scheme of screening.

A screening program for streptomycin sulfate for injection may operate as follows:

(1)   Samples are assayed initially to obtain a single potency estimate for each using a 2 + 2 assay design having a precision of $\pm 7\%$ ($P = 0.95$).

(2)   Samples with an estimated potency on the "as is" basis of 720 IU/mg or more are accepted as satisfactory without further investigation.

(3)   Samples with an estimated potency on the "as is" basis of less than 720 IU/mg are reassayed using the same 2 + 2 assay design of the same precision on one or if necessary two more occasions.

(4)   If the simple mean of the two (or three) results is consistent with the range mean test, and if the mean estimated potency is 720 IU/mg or more, then the sample is accepted as being satisfactory.

(5)   If the simple mean of three results as determined above,is less than than 720 IU/mg, then the assay is repeated (moisture content being determined) so as to obtain a result of sufficient precision and accuracy to enable the analyst to issue his report with confidence and in accordance with the legally defined standards pertaining.

Naturally, in all cases when an adverse report is to be issued, the directions of the pharmacopeia as given in "General Notices" and individual monographs must be observed.

The rationale of the scheme described for screening of streptomycin injection is that for an assay of this assumed precision ($\pm 7\%$ at $P = 0.95$), of

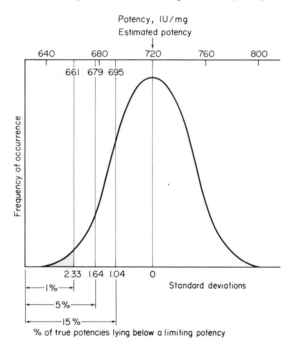

Fig. 9.1   An illustration of the rationale of the screening program outlined in Section 9.6, in which the arbitrary retest limit is set at 720 IU/mg. The graph refers to all single assays leading to an estimated potency of exactly 720 IU/mg with confidence limits of $\pm 7\%$ (i.e., ± 50 IU/mg) at $P = 0.95$. The ordinates indicate the probable relative frequency of the true potency having any specified value. It follows that the probability of the true potency being below any specified value is indicated by the proportion of the total area under the curve that lies to the left of the vertical line for that potency. Thus, the shaded area to the left of the vertical line for potency 661 IU/mg corresponds to 1% of the whole area.

those samples giving estimated potencies of exactly 720 IU/mg, probabilities of true potencies being any particular value can be defined.

Thus:

15% of true potencies would probably be less than 695 IU/mg,
5% of true potencies would probably be less than 679 IU/mg,
1% of true potencies would probably be less than 661 IU/mg.

Allowing for the assumed 3% moisture content, these figures correspond to potencies on the dried basis of 718, 700, and 681 IU/mg, respectively. Such values can be calculated simply from statistical tables. The principle is shown graphically in Fig. 9.1.

In conclusion, all testing involves an element of chance (however small) that the sample is unrepresentative or that the result is significantly in error. While perfection may be rarely attainable, deployment of testing facilities can be so planned as to attain maximum benefit from limited resources.

References

Fisher, R. A., and Yates, F. (1963). "Statistical Tables for Use in Biological, Agricultural and Medical Research." Longman, London.

Humphrey, J. H., Mussett, M. V., and Perry, W. L. M. (1952). The Second International Standard for Penicillin, WHO/BS/170.18.IX.1952.

Lees, K. A., and Tootill, J. P. R. (1955). *Analyst* **80**, 95.

Lightbown, J. W. (1973). In Sampling and Analytical Control. Report on Edinburgh Seminar. Secretariat of the European Free Trade Association.

Perry, W. L. M. (1950). M. R. C. Spec. Rep. Ser. No. 270.

Simpson, J. S. (1963). *In* "Analytical Microbiology" (F. W. Kavanagh, ed.), Vol. I. Academic Press, New York.

Turkey (1969). Güner, Ü., Hewitt, W., İnak, M. and Önal, Ü. Türk. Hijiyen ve Tecrübi Biyoloji Dergisi.

APPENDIX 1

PATTERNS FOR SMALL PLATE ASSAYS

(A) Patterns suitable for use with 2 + 2 or 3 + 3 assays (Figure A1.1). There is no restriction on the manner of allocation of position numbers to treatments. However, the following convention is a convenient routine:

2 + 2 Assays

	Sample 1	Sample 2	Standard
High dose	1	3	5
Low dose	2	4	6

3 + 3 Assays

	Sample	Standard
High dose	1	4
Medium dose	2	5
Low dose	3	6

(B) Pattern used for the 5 + 1 assay (Figure A1.2). Position R is occupied by the reference dose, which is the midpoint of the standard curve. Positions S are occupied by any one of the other points of the standard curve or any one of the samples.

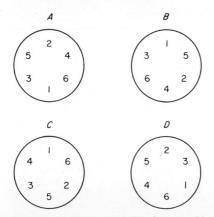

Fig. A1.1. Patterns suitable for use with 2 + 2 or 3 + 3 assays.

Fig. A1.2. Pattern used in the 5 + 1 assay.

CALCULATIONS FOR PARALLEL LINE ASSAYS

Table A2.1

Computation of E and F for Balanced Designs

Number of dose levels	Number of samples	E
2	1	$\frac{1}{2}[(S_2 + T_2) - (S_1 + T_1)]$
	2	$\frac{1}{3}[(S_2 + T_2{}^1 + T_2{}^2) - (S_1 + T_1{}^1 + T_1{}^2)]$
	3	$\frac{1}{4}[(S_2 + T_2{}^1 + T_2{}^2 + T_2{}^3) - (S_1 + T_1{}^1 + T_1{}^2 + T_1{}^3)]$
	7	$\frac{1}{8}[(S_2 + T_2{}^1 + \cdots + T_2{}^7) - (S_1 + T_1{}^1 + \cdots + T_1{}^7)]$
3	1	$\frac{1}{4}[(S_3 + T_3) - (S_1 + T_1)]$
	2	$\frac{1}{6}[(S_3 + T_3{}^1 + T_3{}^2) - (S_1 + T_1{}^1 + T_1{}^2)]$
4	1	$\frac{1}{20}[3(S_4 + T_4) + (S_3 + T_3) - (S_2 + T_2) - 3(S_1 + T_1)]$

Table A2.2

Computation of E and F for Unbalanced Designs

Number of standard dose levels	Number of sample dose levels	Number of samples	E
2	1	i	$S_2 - S_1$
3	2	1	$\frac{1}{5}[(2S_3 + T_2) - (2S_1 + T_1)]$
5	1	i	$\frac{1}{10}[2S_{5C} + S_{4C} - S_{2C} - 2S_{1C}]$
4	1	12	$\frac{1}{10}[3S_4 + S_3 - S_2 - 3S_1]$

[a] Dose levels must be in the same geometrical progression for standard and sample.

[b] S_1, S_2, S_3, etc., refer to mean responses to dose levels of standard increasing in geometric progression.

[c] T_1, T_2, T_3, etc., refer to mean responses to dose levels of sample increasing in geometric progression.

[d] Superscripts as in $T_2{}^1$, $T_2{}^2$, $T_2{}^i$, etc., distinguish different samples.

F	Suitable assay design	Notes
$\frac{1}{2}[(T_2 + T_1) - (S_2 + S_1)]$	Petri dish or 4 × 4 Latin square	b, c
$\frac{1}{2}[(T_2{}^i + T_1{}^i) - (S_2 + S_1)]$	Petri dish or 6 × 6 Latin square	b, c, d
$\frac{1}{2}[(T_2{}^i + T_1{}^i) - (S_2 + S_1)]$	8 × 8 Latin square	b, c, d
$\frac{1}{2}[(T_2{}^i + T_1{}^i) - (S_2 + S_1)]$	8 × 8 quasi-Latin square	b, c, d
$\frac{1}{3}[(T_3 + T_2 + T_1) - (S_3 + S_2 + S_1)]$	Petri dish or 6 × 6 Latin square	a, b, c
$\frac{1}{3}[(T_3{}^i + T_2{}^i + T_1{}^i) - (S_3 + S_2 + S_1)]$	9 × 9 Latin square	a, b, c, d
$\frac{1}{4}[(T_4 + T_3 + T_2 + T_1) - (S_4 + S_3 + S_2 + S_1)]$	8 × 8 Latin square	a, b, c

F	Suitable assay design	Notes
$T^i - S_2$	Petri dish	b, d, e
$\frac{1}{2}[T_1 - T_2] - \frac{1}{3}[S_1 + S_2 + S_3]$	Petri dish	a, b, c, g
$T^i - S_3{}^i$	Petri dish	b, d, f
$T^i - \frac{1}{4}[S_1 + S_2 + S_3 + S_4]$	Large plate 8 × 8 random distribution	b, d, g

[e] This may be arranged *either* for four samples on each plate as a self-contained assay *or* for any number of samples using plate corrections as described in Section 2.14.

[f] Additional subscripts such as in S_{1C}, S_{2C}, etc., indicate that the mean responses S_1, S_2 have been modified by the plate correction as described in Section 2.14.

[g] In this assay $\bar{x}_S$ and $\bar{x}_T$ are normally unequal. For calculation of log potency ratio in this special case see Section 2.13.

APPENDIX 3

POTENCY RATIO, OR *F/E*, TABLES

These tables were introduced in Section 2.7, where their method of use is also described.

Table A3.1

2:1 Dose Ratio Assays

F/E	R	F/E	R	F/E	R
−1.000	0.50	−0.252	0.84	+0.239	1.18
−0.971	0.51	−0.235	0.85	+0.251	1.19
−0.943	0.52	−0.218	0.86	+0.263	1.20
−0.916	0.53	−0.201	0.87	+0.275	1.21
−0.889	0.54	−0.184	0.88	+0.287	1.22
−0.863	0.55	−0.168	0.89	+0.299	1.23
−0.837	0.56	−0.152	0.90	+0.310	1.24
−0.811	0.57	−0.136	0.91	+0.322	1.25
−0.786	0.58	−0.120	0.92	+0.333	1.26
−0.761	0.59	−0.105	0.93	+0.345	1.27
−0.737	0.60	−0.089	0.94	+0.356	1.28
−0.713	0.61	−0.074	0.95	+0.367	1.29
−0.690	0.62	−0.059	0.96	+0.379	1.30
−0.667	0.63	−0.044	0.97	+0.390	1.31
−0.644	0.64	−0.029	0.98	+0.401	1.32
−0.622	0.65	−0.015	0.99	+0.411	1.33
−0.600	0.66	0.000	1.00	+0.422	1.34
−0.578	0.67	+0.014	1.01	+0.433	1.35
−0.556	0.68	+0.029	1.02	+0.444	1.36
−0.535	0.69	+0.043	1.03	+0.454	1.37
−0.515	0.70	+0.057	1.04	+0.465	1.38
−0.494	0.71	+0.070	1.05	+0.475	1.39
−0.474	0.72	+0.084	1.06	+0.485	1.40
−0.454	0.73	+0.098	1.07	+0.496	1.41
−0.434	0.74	+0.111	1.08	+0.506	1.42
−0.415	0.75	+0.124	1.09	+0.516	1.43
−0.396	0.76	+0.138	1.10	+0.526	1.44
−0.377	0.77	+0.151	1.11	+0.536	1.45
−0.358	0.78	+0.164	1.12	+0.546	1.46
−0.340	0.79	+0.176	1.13	+0.556	1.47
−0.322	0.80	+0.189	1.14	+0.566	1.48
−0.304	0.81	+0.202	1.15	+0.575	1.49
−0.286	0.82	+0.214	1.16	+0.585	1.50
−0.269	0.83	+0.227	1.17		

Table A3.2

Extended Table for 2:1 Dose Ratio Low Precision Assays

F/E	R	F/E	R	F/E	R	F/E	R
−4.32	0.05	+0.765	1.70	+1.378	2.60	+2.46	5.50
−3.32	0.10	+0.807	1.75	+1.433	2.70	+2.59	6.00
−2.74	0.15	+0.848	1.80	+1.486	2.80	+2.70	6.50
−2.32	0.20	+0.888	1.85	+1.536	2.90	+2.81	7.00
−2.00	0.25	+0.926	1.90	+1.585	3.00	+2.91	7.50
−1.74	0.30	+0.963	1.95	+1.678	3.20	+3.00	8.00
−1.52	0.35	+1.000	2.00	+1.766	3.40	+3.09	8.50
−1.32	0.40	+1.070	2.10	+1.848	3.60	+3.17	9.00
−1.15	0.45	+1.137	2.20	+1.926	3.80	+3.25	9.50
+0.632	1.55	+1.201	2.30	+2.00	4.00	+3.32	10.00
+0.678	1.60	+1.263	2.40	+2.17	4.50		
+0.723	1.65	+1.322	2.50	+2.32	5.00		

Table A3.3

4:1 Dose Ratio Assays

F/E	R	F/E	R	F/E	R	F/E	R
−0.500	0.50	−0.198	0.76	+0.014	1.02	+0.178	1.28
−0.486	0.51	−0.189	0.77	+0.021	1.03	+0.184	1.29
−0.472	0.52	−0.179	0.78	+0.028	1.04	+0.189	1.30
−0.458	0.53	−0.170	0.79	+0.035	1.05	+0.195	1.31
−0.445	0.54	−0.161	0.80	+0.042	1.06	+0.200	1.32
−0.431	0.55	−0.152	0.81	+0.049	1.07	+0.206	1.33
−0.418	0.56	−0.143	0.82	+0.056	1.08	+0.211	1.34
−0.406	0.57	−0.134	0.83	+0.062	1.09	+0.217	1.35
−0.393	0.58	−0.126	0.84	+0.069	1.10	+0.222	1.36
−0.381	0.59	−0.117	0.85	+0.075	1.11	+0.227	1.37
−0.368	0.60	−0.109	0.86	+0.082	1.12	+0.232	1.38
−0.357	0.61	−0.101	0.87	+0.088	1.13	+0.238	1.39
−0.345	0.62	−0.092	0.88	+0.095	1.14	+0.243	1.40
−0.333	0.63	−0.084	0.89	+0.101	1.15	+0.248	1.41
−0.322	0.64	−0.076	0.90	+0.107	1.16	+0.253	1.42
−0.311	0.65	−0.068	0.91	+0.113	1.17	+0.258	1.43
−0.300	0.66	−0.060	0.92	+0.119	1.18	+0.263	1.44
−0.289	0.67	−0.052	0.93	+0.126	1.19	+0.268	1.45
−0.278	0.68	−0.045	0.94	+0.132	1.20	+0.273	1.46
−0.268	0.69	−0.037	0.95	+0.138	1.21	+0.278	1.47
−0.257	0.70	−0.029	0.96	+0.143	1.22	+0.283	1.48
−0.247	0.71	−0.022	0.97	+0.149	1.23	+0.288	1.49
−0.237	0.72	−0.015	0.98	+0.155	1.24	+0.293	1.50
−0.227	0.73	−0.007	0.99	+0.161	1.25		
−0.217	0.74	0.000	1.00	+0.167	1.26		
−0.207	0.75	+0.007	1.01	+0.172	1.27		

PROFORMA AND WORKED EXAMPLE— AGAR DIFFUSION ASSAY

<u>MICROBIOLOGICAL ASSAY</u>

Design: 2 + 2 using either
6 x 6 Latin square or Petri dish.
Standard high dose: 80 $^{IU}/_{ml}$ Dose ratio: $4:1$

Date: $18/6/72$
Substance: $streptomycin$
Test organism: $B.\ subtilis$

Solution no.	Sample 1		Sample 2		Standard			
	1	2	3	4	5	6		
	20.1	15.7	20.3	15.9	19.8	15.3		
	20.9	16.5	20.5	15.8	20.7	15.9		
	20.9	16.4	20.5	15.9	20.4	16.6		
	20.8	16.7	20.2	16.2	21.0	16.3		
	20.6	16.8	20.5	15.8	20.2	16.4		
	19.9	16.5	20.1	15.7	20.3	15.8		
High dose totals	123.2		121.1		122.4		367.7	S(H)
Low dose totals		98.6		95.3		96.3	290.2	S(L)
Sample totals S(T)	221.8		217.4		218.7		Standard total S(S)	
D = S(T) − S(S)	+3.1		−1.3		———			

Calculate B = S(H) − S(L) = $367.7 - 290.2 = 77.5$

then   2B/3 = (2 x 77.5)/3   = 51.67

Obtain relative potencies from the chart for 4 :1 dose ratio

using the relationship $\dfrac{D}{2B/3} = \dfrac{F}{E}$

For sample (1)   F/E = $+3.1/51.67$ = $+0.060$, giving R = 1.09

For sample (2)   F'/E = $-1.3/51.67$ = -0.025 , giving R' = 0.97

SOURCES OF REFERENCE MATERIALS

A5.1. International Biological Standards and International Biological Reference Preparations

About 40 of these antibiotic substances are held and distributed for the World Health Organization by International Laboratory for Biological Standards, National Institute for Biological Standardisation and Control, Holly Hill, Hampstead, London, England.

Samples are distributed free of charge to national laboratories for biological standards, as well as to other biological laboratories in countries where national laboratories for biological standards do not function.

A5.2. British Biological Standards and Reference Preparations

These are generally identical with the corresponding international preparation. Additionally there is (1976) a small number of substances for which no corresponding international preparation has been established. They are distributed from the same London address as the international preparations described above.

A5.3. European Pharmacopeia Commission Reference Substances

These include a small number of antibiotics, some of which are intended as chemical reference substances and others as biological reference substances.

They are obtainable from Secretariat of the European Pharmacopeia Commission, Council of Europe, 67000 Strasbourg, France.

A5.4. United States Pharmacopeia Reference Substances

This extensive list of pharmaceutical substances includes about 30 antibiotics and 10 vitamins. Some of these are intended as chemical reference substances and others as biological reference substances.

They are distributed by USP Reference Standards, 4630 Montgomery Avenue, Bethesda, Maryland 20014.

A5.5.   International Chemical Reference Substances

These pharmaceutical substances are intended primarily for use in infrared identification, chromatographic tests and assays, and in spectrophotometric or photometric methods of assay. However, the International Pharmacopeia prescribes the use of a few of these chemical reference substances in the biological assay of pharmaceutical dosage forms of some of the semisynthetic penicillins.

They are distributed by WHO International Reference Centre for Chemical Reference Substances, Apotekens Centrallaboratorium, Box 333, Solna 3, Sweden.

THE DILUTION OF REFERENCE STANDARDS

Convenient dilution schemes are suggested in this appendix. The dilutions apply primarily to the potency ranges in which lie all antibiotics in regular current use.

Depending on the stability of the substance in solution, it may be possible to retain the primary dilution as a master solution, storing it in a refrigerator for a week or more and making further dilutions each day as required. In this case, the master solution should be allowed to reach room temperature before pipeting for greater accuracy.

Weighings should be within about $\pm 5\%$ of the figures suggested here but should be recorded for calculation purposes to the nearest 0.1 mg. The weights suggested have been kept reasonably small so as to economize in the consumption of reference material and to avoid the necessity of extensive dilution. Weights are not so small, however, as to require a semimicrobalance (sensitivity 0.01 mg).

Dilutions avoid the use of pipets smaller than 5 ml. The manufacturer's percentage tolerances are wider on the smaller pipettes; this, together with probable greater errors in their use, constitutes an avoidable source of error in the assay result. All the dilutions tabulated lead to a *nominal* potency of 20 U/ml. It is unnecessary to work with exactly round figures. Indeed, it may be a disadvantage to do so either because

(1)   the exactly required amount is calculated and weighed—a tedious business permitting time for significant moisture pick up on a small quantity of a hygroscopic substance, or

(2)   the approximately required amount is weighed, recorded exactly, and the exactly required dilution calculated. This necessitates dilution to awkward volumes such as are shown in Examples 2 and 8.

Dilution Scheme

(1)   Dilutions to 20 U/ml:

Approximate standard potency, U/mg	Primary dilution		Secondary dilution	
	mg	ml	ml	ml
50	100 $\longrightarrow$ 250			
60	83 $\longrightarrow$ 250			
70	72 $\longrightarrow$ 250			
80	63 $\longrightarrow$ 250			
600	67 $\longrightarrow$ 100	:	5 $\longrightarrow$ 100	

Approximate standard potency, U/mg	Primary dilution		Secondary dilution	
	mg	ml	ml	ml
700	57 → 100	:	5 → 100	
800	50 → 100	:	5 → 100	
900	88 → 200	:	5 → 100	
1000	80 → 200	:	5 → 100	
1100	72 → 200	:	5 → 100	
1600⎱ 1700⎰	50 → 200	:	5 → 100	
3000⎱ 3200⎰	50 → 100	:	5 → 200	

(2)   Dilutions to less than 20 U/ml: When the final tabulated dilution is 5 ml → 100 ml, it may be replaced by 5 ml → 200 ml, giving 10 U/ml, or 5 ml → 250 ml, giving 8 U/ml.

Further dilutions of the 20 U/ml solutions may be made thus:

Tertiary dilution, ml	Resultant potency, U/ml
25 → 100	5
10 → 50	4
5 → 50	2
5 → 100	1

It is convenient to base calculations on the nominal potency of the high dose test solution and apply a factor just as is done routinely in volumetric chemical assays, e.g., neomycin sulfate, potency 672 IU/mg:

required test solution potencies: 40 IU/ml   high dose
                                                    10 IU/ml   low dose

Replace the tabulated secondary dilution by 10 ml → 100 ml, so as to obtain 40 IU/ml in place of 20 IU/ml. Supposing the actual weight of neomycin sulfate were 58.2 mg, then the dilutions

$$58.2 \quad \text{mg} \to 100 \quad \text{ml} : 10 \quad \text{ml} \to 100 \quad \text{ml}$$

lead to a potency for the high dose test solution of

$$\frac{58.2 \times 672 \times 10}{100 \times 100} = 39.1 \quad \text{IU/ml}$$

This solution may be described as being of nominal potency 40 IU/ml but having a factor of 39.1/40 = 0.978.

THE PROBIT TRANSFORMATION

Probits were introduced in Section 4.3.

A. 1–99%

Percentages ⟶ 0	1	2	3	4	5	6	7	8	9
0		2.67	2.95	3.12	3.25	3.36	3.45	3.52	3.59	3.66
10	3.72	3.77	3.82	3.87	3.92	3.96	4.01	4.05	4.08	4.12
20	4.16	4.19	4.23	4.26	4.29	4.33	4.36	43.9	4.42	4.45
30	4.48	4.50	4.53	4.56	4.59	4.61	4.64	4.67	4.69	4.72
40	4.75	4.77	4.80	4.82	4.85	4.87	4.90	4.92	4.95	4.97
50	5.00	5.03	5.05	5.08	5.10	5.13	5.15	5.18	5.20	5.23
60	5.25	5.28	5.31	5.33	5.36	5.39	5.41	5.44	5.47	5.50
70	5.52	5.55	5.58	5.61	5.64	5.67	5.71	5.74	5.77	5.81
80	5.84	5.88	5.92	5.95	5.99	6.04	6.08	6.13	6.18	6.23
90	6.28	6.34	6.41	6.48	6.55	6.64	6.75	6.88	7.05	7.33

B. 99–99.9%

Percentages ⟶ 0.0	0.1	0.2	0.3	0.4	0.5	0.6	0.7	0.8	0.9
99	7.33	7.37	7.41	7.46	7.51	7.58	7.65	7.75	7.88	8.09

APPENDIX 8

THE ANGULAR TRANSFORMATION

The angular transformation is defined by the equation

$$p = \sin^2 \phi \qquad (\text{Eq. (4.8)})$$

which was introduced in Section 4.3. Values of the angle ϕ related to p, the proportionate (percent) response by this equation, are tabulated.

$p\%$ →	0	2	4	6	8
0		8.1	11.5	14.2	16.4
10	18.4	20.3	22.0	23.6	25.1
20	26.6	28.0	29.3	30.7	32.0
30	33.2	34.5	35.7	36.9	38.1
40	39.2	40.4	41.6	42.7	43.9
50	45.0	46.2	47.3	48.5	40.6
60	50.8	51.9	53.1	54.3	55.6
70	56.8	58.0	59.3	60.7	62.0
80	63.4	64.9	66.4	68.6	69.7
90	71.6	73.6	75.8	78.5	81.9

APPENDIX 9

THE t DISTRIBUTION

"Student's t" was introduced in Section 6.5. This distribution relates probability and degrees of freedom. At high values for degrees of freedom it approaches the normal distribution. The values of t tabulated here refer to $P = 0.95$ only.

Degrees of freedom	t	Degrees of freedom	t
1	12.71	11	2.20
2	4.30	12	2.18
3	3.18	15	2.13
4	2.78	20	2.09
5	2.57	25	2.06
6	2.45	30	2.04
7	2.37	40	2.02
8	2.31	60	2.00
9	2.26	120	1.98
10	2.23	∞	1.96

APPENDIX 10

VARIANCE RATIO TABLES—THE F TEST

The F test for significance of variance ratio was introduced in Section 6.4. The variance ratio F may be defined as the ratio of the two mean squares that are to be compared. In all cases the larger, which has n_1 degrees of freedom, is the numerator and the smaller, which has n_2 degrees of freedom, is the denominator. The calculated value of F is compared with the appropriate tabulated values in the tables for various probabilities. Interpretation of the comparisons is illustrated in Section 6.4.

Table A10.1

1% Probability

n_2 \ n_1	1	2	3	4	5
1	4052	4999	5403	5625	5764
2	98.5	99.0	99.2	99.3	99.3
3	34.1	30.8	29.5	28.7	28.2
4	21.2	18.0	16.7	16.0	15.5
5	16.3	13.3	12.1	11.4	11.0
10	10.0	7.6	6.6	6.0	5.6
15	8.7	6.4	5.4	4.9	4.6
20	8.1	5.9	4.9	4.4	4.1
30	7.6	5.4	4.5	4.0	3.7

Table A10.2

5% Probability

n_2 \ n_1	1	2	3	4	5
1	161.4	199.5	215.7	224.6	230.2
2	18.5	19.0	19.2	19.3	19.3
3	10.1	9.6	9.3	9.1	9.0
4	7.7	6.9	6.6	6.4	6.3
5	6.6	5.8	5.4	5.2	5.0
10	5.0	4.1	3.7	3.5	3.3
15	4.5	3.7	3.3	3.1	2.9
20	4.4	3.5	3.1	2.9	2.7
30	4.2	3.3	2.9	2.7	2.5

Table A10.3

10% Probability

n_2 \ n_1	1	2	3	4	5
1	39.9	49.5	53.6	55.8	57.2
2	8.5	9.0	9.2	9.2	9.3
3	5.5	5.5	5.4	5.3	5.3
4	4.5	4.3	4.2	4.1	4.0
5	4.1	3.8	3.6	3.5	3.5
10	3.3	2.9	2.7	2.6	2.5
15	3.1	2.7	2.5	2.4	2.3
20	3.0	2.6	2.4	2.3	2.2
30	2.9	2.5	2.3	2.1	2.1

Table A10.4

20% Probability

n_2 \ n_1	1	2	3	4	5
1	9.47	12.00	13.06	13.64	14.01
2	3.56	4.00	4.16	4.24	4.28
3	2.68	2.89	2.94	2.96	2.97
4	2.35	2.47	2.48	2.48	2.48
5	2.18	2.26	2.25	2.24	2.23
10	1.88	1.90	1.86	1.83	1.80
15	1.80	1.79	1.75	1.71	1.68
20	1.76	1.75	1.70	1.65	1.62
30	1.72	1.70	1.64	1.60	1.57

THE χ^2 DISTRIBUTION[a]

n \ P	0.20	0.10	0.05	0.01
1	1.64	2.71	3.84	6.64
2	3.22	4.61	5.99	9.21
3	4.64	6.25	7.82	11.35
4	5.99	7.78	9.94	13.28
5	7.29	9.24	11.07	15.09
6	8.56	10.65	12.59	16.81
7	9.80	12.02	14.07	18.48
8	11.03	13.36	15.51	20.09

[a] The χ^2 distribution was introduced in Section 9.3.

APPENDIX 12

THE RANGE/MEAN TEST[a]

Number of assays	Assumed standard error							
	1%	2%	3%	4%	5%	6%	7%	8%
2	0.028	0.055	0.083	0.110	0.138	0.165	0.192	0.220
3	0.034	0.067	0.100	0.134	0.164	0.200	0.234	0.268
4	0.036	0.073	0.109	0.145	0.180	0.218	0.254	0.290
5	0.038	0.076	0.115	0.153	0.190	0.230	0.268	0.306
6	0.040	0.080	0.121	0.161	0.200	0.242	0.282	0.322
7	0.042	0.084	0.125	0.168	0.210	0.252	0.294	0.336
8	0.044	0.087	0.130	0.174	0.219	0.261	0.304	0.348

[a] The range/mean test was introduced in Section 9.4, where use of the range/mean table is also described.

EVIDENCE THAT QUADRATIC CURVATURE IS WITHOUT INFLUENCE IN BALANCED PARALLEL LINE ASSAYS

Standard expressions for calculation of potency estimate in parallel line assays have been demonstrated in Chapter 2 and summarized in Appendix 2. Basic assumptions in the derivation of these expressions include the following: (i) log dose–response lines are straight, (ii) log dose–response lines for standard and unknown sample are parallel.

It is demonstrated here (by means of the 3 + 3 design) that a quadratic component of curvature is entirely without influence on the estimated potency.

Suppose that:

(1) Response (y) is related to log dose (x) by an equation of the form

$$y = \alpha + \beta x + \gamma x^2 \qquad (A13.1)$$

in which γx^2 is the component of quadratic curvature.

(2) A standard preparation has three dose levels such that

$$x = -I, \quad 0, \quad \text{and} \quad +I$$

(3) An unknown preparation has three dose levels spaced as for the standard but R times as potent; if $\log R = Q$, then the corresponding values of x for the unknown preparation will be $Q - I$, Q, and $Q + I$.

Using the nomenclature of the International Pharmacopeia, mean responses to the six doses are $S_1, S_2, S_3, T_1, T_2,$ and T_3. Expressions for these in terms of the six values of x are derived from Eq. (A13.1) and shown in tabular form:

x	y
$-I$	$\alpha - \beta I + \gamma I^2 = S_1$
0	$\alpha = S_2$
$+I$	$\alpha + \beta I + \gamma I^2 = S_3$
$Q - I$	$\alpha + \beta(Q - I) + \gamma(Q^2 - 2QI + I^2) = T_1$
$+Q$	$\alpha + \beta Q + \gamma Q^2 = T_2$
$Q + I$	$\alpha + \beta(Q + I) + \gamma(Q^2 + 2QI + I^2) = T_3$

Working out these expression and substituting in the standard forms for evaluation of E and F in a $3 + 3$ assay (Appendix 2), we get

$$E = \tfrac{1}{4}[(2\alpha + \beta Q + 2\beta I + \gamma Q^2 + 2\gamma I^2 + 2\gamma QI)$$
$$- (2\alpha + \beta Q - 2\beta I + \gamma Q^2 + 2\gamma I^2 - 2\gamma QI)] \qquad (A13.2)$$

which simplifies to

$$E = I[\beta + \gamma Q]$$

Also,

$$F = \tfrac{1}{3}[(3\alpha + 3\beta Q + \beta I - \beta I + 3\gamma Q^2 + 2\gamma I^2 + 2QI - 2QI)$$
$$- (3\alpha + \beta I - \beta I + 2\gamma I^2)] \qquad (A13.3)$$

which simplifies to

$$F = Q[\beta + \gamma Q]$$

Again using the nomenclature of the International Pharmacopeia,

$$b = E/I \qquad (A13.4)$$

so that

$$b = I[\beta + \gamma Q]/I = \beta + \gamma Q \qquad \text{and} \qquad M = F/b \qquad (A13.5)$$

so that

$$M = Q[\beta + \gamma Q]/[\beta + \gamma Q] = Q$$

Thus, the *estimated* log of potency ratio M is identical with the *known* log of potency ratio Q.

It follows that the component γx^2 of Eq. (A13.1) has not invalidated the assumptions implicit in the use of expressions (A13.2)–(A13.5).

INDEX

A

Absorptiometry in tube assays, 70, 72–74
Agar diffusion assay, 17–68, *see also* Parallel
    line assays
  basic techniques and principles, 2, 3
  inoculum level, 22
  log dose–response curves, 27–30
  thickness of agar, 22–24.
  tubes, 26
Angular transformation, 110, 112, 113, 116,
    120–123, 127–130, 270
  parallel line assays evaluation, 194–202
  slope ratio assays evaluation, 210–213
  table, 270
Anhydrotetracycline interference in
    tetracycline assays, 215
Antibiotic activity, total measurement, 136,
    138
Antibiotics
  bactericidal, 142
  bacteriostatic, 142
  solubilities, 139
Antibiotic tube assays
  linearization of dose–response
    relationships, theoretical
    considerations, 107–113
  log response versus dose relationship,
    130, 131
Automation, 5–7
Autoturb®, 6, 132, 133, 134, 135.

B

Balanced designs, 27, 31, 97–101, 120, 219,
    221, 222
  comparison of $2 + 2$ with $3 + 3$, 221, 222
  inherent advantage, 219
  multiple slope ratio assay, 97–101
  for parallel line assays, 27, 31, 120
  unbalanced designs and, 221

Bioautograph, 142, 145–147, 191–194
  qualitative, 142
  quantitative, 145–147, 191–194
    evaluation, 191–194
    potency estimation, 145–147
Biological activity unit, 12

C

Cell concentration, relative, 106
Chi squared
  approximate test, 247–251
  table, 274
Chloramphenicol, esters of, 139
Chloramphenicol palmitate, 139
Chloramphenicol sodium succinate, 139
Choice of method and design, 214–242
  choice of method, 214–216
Chromatography
  bioautograph techniques
    antibiotic E129, 140
    comparative bioautographs, 142–145
    gentamycin complex, 140
    penicillins, 140
    penicillin V, 140
    quantitative bioautographs, 145–147
    vitamin B_{12}, 140, 142
  paper, of gentamycin complex, 140
Coded log doses, convenience of, 45
Collaborative assays, 40, 245, 246
  international, 11
  moisture content of preparations, 14
Computer interpolation from standard
    curves, 131, 133, 134
Confidence limits, 152, 153, 165, 167, 182,
    206
  alternative method of estimating, 167, 182
  of potency estimate by parallel line assays,
    152, 153
    general expression, 165
  of potency estimate by slope ratio assay,
    general expression, 206

Critical concentration, 19, 20, 22, 26
Critical population, 19–21
Critical time, 19–21, 26
Curvature, 26, 27, 39–43, 70, 71, 113, 159,
    163, 183, 225–228, 276, 277
  biases arising from in parallel line assays,
    226–228
  check by three dose level design, 39–43
  cubic, 159, 163, 183
  of dose–response line, slope ratio assays,
    70, 71
  influence on parallel line assays, 225–228
  of log dose–response line, 26, 27 39–43,
    226–228
  quadratic, 27, 159, 163, 183, 228, 276, 277
  lack of influence in balanced parallel line
    assays, 228, 276, 277
  of response line in tube assays for
    antibiotics, linearization procedures,
    113

D

Degrees of freedom, 160
Design of assay, selection of
  general considerations, 216, 217
  plate assay designs, 217–225, 228–232
  slope ratio designs, 233–238, 240–241
  tube assays for antibiotics, 241, 242
Diffusion time, influence on zone size, 49

E

Economical tests, 229
Edge effects in large plate assays, 51, 52
Effective constituent, 5
Efficiency in assay, 132
Electrophoresis of antibiotics, 140
Errors, sources of, 13–15, 74
  moisture content, 13
  optical measurements, 74
  volumetric measurements, 14
  weighings, 13
  weight variation of unit dosage forms, 15
Evaluation
  of parallel line assays, 151–202.
  of slope ratio assays, 203–213

F

F test, 160, *see also* Variance ratio
Factorial calculations, 47, 48
Factorial tables, 107, 260, 261
Five dose level standard curve
  arithmetical estimation of potency, 65
  graphical estimation of potency, 65
  tube assays for antibiotics, 106
Five point common zero assay, 241

G

Geometrical progression of dose levels
  in parallel line assays, 40
  in tube assays for antibiotics, 106
Generation time, 21, 108, 109
Growth-inhibiting substances, 4
Growth inhibition, mechanisms, 108
Growth-promoting substances, 4, 5, 70–71
  principles of assay, 4, 5
  tube assays for, 70–101.

H

Heterogeneity
  of potency estimates, correction factor,
    248–251
  of samples, 244
Heterogeneous materials biological assay,
  138
Homogeneity of potency estimates, test for,
  247–251
Homoscedasticity, 44

I

Incubation, 70, 71, 104
  period of
    in tube assays for antibiotics, 104
    in tube assays for growth-promoting
      substances, 70, 71
  temperature of
    in tube assays for antibiotics, 104
    in tube assays for growth-promoting
      substances, 71

termination of in tube assays for antibiotics, 104
Inhibiting substances, interferences from, 71, 103
Inhibition zones, theory of formation, 17–26
  double, 19
  influences on size, 24
Inhibitory population, 19, 23
Inoculum
  population, 19–24
  in tube assays for growth-promoting substances, 71
Interferences in assays, detection, 40
Inverse log plot, 113
Iterative calculations, 67

L

Lag phase of growth, 21, 105, 108
  in tube assays for antibiotics, 105
Large plates, 49–52, 173–187, 219
Large plate assays
  evaluation
    Latin square design, 173–177
    low-precision random design, 183–187
    quasi-Latin square design, 177–182
  general principles, 49–52
  Latin square designs, 49–54, 219, 230–232
    4 × 4, 52
    6 × 6, 49, 52
    8 × 8, 49–54
    9 × 9, 52
    12 × 12, 52, 232
Logarithmic phase of growth in tube assays for antibiotics, 105
Log potency ratio, *M*
  standard error, 166
  variance of, 166
Logit transformation, 110–113
Low-precision assays using large plates, 57–61.

M

Matrix equations, 211
Mechanization, 5–7
Median response, 105
Microgram equivalent, 12

Missing values
  influence on statistical evaluation, 67
  replacement of, 66–68
Mixtures of antibiotics, 136–149
  general assay techniques
    differential assays, 142, 147–149
    interference thresholds, 141
    selective destruction, 139, 141
    selective determination, 139, 141
    separation techniques, 139
  occurrence, 136–139
Moisture content in specifications, 255
Multiple assay designs
  for parallel line assays, 36–38, 49–57, 321
  for slope ratio assays, 97–101

N

Neomycin, problems of assay, 136, 177
Nephelometry in tube assays, 70, 72, 73, 112
Nicotinic acid, dose–response curve, 239
Nonideal responses in tube assays for growth-promoting substances, 74–81.
  curve-straightening procedure, 75–81
Normal distribution curve, 110, 256
Normal equivalent deviation, 110

O

Opposed curvature, 171
Optical measurements, flow birefringence, 72
Orthogonal polynomial coefficients, 156, 158, 163, 164, 171, 190, 192, 200

P

Parallel line assays, 30–68, 127–130, 145–147, 167–182, 188–191, 194–197, 222–228, 259–264
  curvature, influence of, 225–228
  E and *F*, computation tables, 260, 261
  5 + 1 (F.D.A.) design, evaluation of, 188–191
  four dose level, 43–48, 58, 127, 128
  interpolation from standard curve, 61–66

Parallel Line Assays (*Cont.*)
  Latin square design, evaluation of, 173–177
  log potency ratio *M*, influence of, 222–225
  multidose level, 44, 58, 61–66, 127–130
  multiple, two dose level assay, evaluation of, 167–170, 173–182
  number of dose levels influence in balanced designs, 222–224
  overall dose range, influence, 223, 224
  potency ratio, computation tables, 262, 263
  proforma for assay observations and computations, 264
  small plate assays, patterns, 259
  quasi-Latin square design, evaluation of, 177–182
  simplification of computation, 38, 39
  single dose level for sample, 66
  six dose level design, 127, 128
  three dose level assay, 39–43
    evaluation of, 170–173
Parallelism, 165
Penicillin, assay in blood, 140
pH in vitamin assays, 74
Plate assays, design choice, 228–232
Plate correction, 62
Plate method, *see* Agar diffusion assay
Precision of potency estimate, 151
Probit transformation, 110, 112–114, 116, 120–123, 125–127, 269
  table, 269
Procaine benzylpenicillin, 138
Purpose of assay, 7, 8
  in relation to design, 7, 216, 217
Purpose built assay designs, 199

Q

Quantal responses, 110
Quasi-Latin square designs, 49–52, 55–57, 178–181, 219, 231
  evaluation of assays, 178–181

R

Randomization, 2, 58, 154
  tables of random numbers, 58
Randomized block design, 67

Range/mean test, 251, 252, 256, 275
  table, 275
Reference standards, 8, 11, 12, 246, 265–268
  for antibiotics, 246
  dilution of, tables, 267, 268
  International Biological Reference Preparations, 8, 265
  International Biological Standards, 8, 265
  International Chemical Reference Substances, 12, 266
  sources, 265, 266
  working standards, 11
Regression equations, 47, 48, 85–90, 92, 93, 159
  multiple linear, 85–90, 92, 93
  simple linear, 47, 48, 85
Regression coefficient
  in parallel line assays, 47, 48
  in slope ratio assays, 86
Repetition of assays, 243–257
  day-to-day variation, 245
Replicate potency estimates, combination of, 247–253.
  simplified methods, 251–253
Replication of treatments, influence on precision
  in parallel line assays, 219, 220, 228
  in slope ratio assays, 232, 234
Reports, analytical, 14–16, 251–257
Reservoirs, alternative forms, 3
Residual error, 152, 160
Response
  in parallel line assays
    standard deviation, 44
    variance, 44
  in tube assays for growth-promoting substances, 72–74

S

Screening
  programs, 255–257
  tests, 142, 229, 230
Significance, 161
Similarity, condition of, 5
Simultaneous equations
  in computation of slope ratio assays, 86, 89
    matrix form, 89

Slope
  estimation in parallel line assays, 45–48
  index of significance, 165, 203
  of log dose–response line, 22, 28–30
    influences on, 22
Slope ratio assays, 81–101, 109, 116, 125,
    126, 130–132, 203–213, 233–241
  arithmetical estimation of potency, 87–90,
    92–96, 99–101
  curvature, influence of, 238–240
  design choice, 240, 241
  design principles, 233–238
  design, critical review, 97
  evaluation, 203–213
    basic principles, 203, 204
    potency estimate, confidence limits, 206
    potency ratio
      standard error, 205
      variance, 203
    validity criteria, 203
  five-point common zero design, 100, 241
  graphical estimation of potency, 83, 91, 98
  multiple assays, 97–101
    evaluation, 207–210
  number of dose levels, influence on
    confidence limits, 234
  omitting response to zero dose, 116
  overall dose range, 233–238
  potency computation
    from balanced design, 90–96
      simplified procedure, 93–96
    from unbalanced design, 87–90
  potency ratio influence on confidence
    limits, 235–238
  replication, influence on confidence limits,
    234
  three-point designs, 132
  unbalanced designs, 81–83
  validity criteria, 81, 83–85
variance of potency ratio, R, influence of
    overall dose range, 233
Specifications, 1, 14–16, 139, 253–257
  manufacturing, 139, 253, 254
  official standards, 254–257
  for pharmaceutical substances, 1, 14–16
Stability testing, 8
Standard preparations, *see* Reference
    standards
Statistical evaluation, 151–213
  limitations, 153

of parallel line assays, 151–202
  basic assumptions, 154
of slope ratio assays, 203–213
  basic assumptions, 203
Student's *t* test, 165, 271
  table, 271
Symmetrical assay designs, *see* Balanced
    designs
Synergism, 145
Synthetic growth media, 4

T

t distribution, 165, 271
Technicon AutoAnalyser®, 7
Temperature, influence on inhibition zones,
    18–26
Test solution preparation, 13, 14, 267, 268
Therapeutic Substances Regulations (Great
    Britain), 254
Thiamine dose–response curve, 239
3 + 2 assay design, 145
Tube assays, 3, 4, 103–107, 194–202, 210,
    232, 241, 242
  for antibiotics
    angular transformation of response, 242
    common forms of representation,
      105–107
    design principles, 241, 242
  evaluation based on function of
      response versus log dose, 194–202
  evaluation as slope ratio assay, 210
  factors influencing response, 103, 104
  probit transformation of response, 242
  slope ratio designs, 242
  basic techniques and principles, 3, 4
  design choice, general considerations, 232
2 + 1 design
  of British and International
      Pharmacopeias, 230
  modification of FDA method, 230
2 + 2 designs, 31–32
  multiple, evaluation, 178

U

Unit of biological activity, 12

V

Validity of assays, 151, 152
Variance analysis, 156, *see also* Variance
    ratio
Variance ratio, 160, 164, 272, 273, *see also*
    Variance analysis
  tables for *F* test, 272, 273
Vitamin assays
  comparison of tube and plate methods,
    215, 216
  specificity, 215, 216

W

Weighted mean of potency estimates, 247
  confidence limits, 248
  variance, 247

Weighting of responses, 44–48, 94, 112
  in parallel line assays, 45
  in slope ratio assays, 94
  weighting coefficients, 112
Working standards, 247

Z

Zone diameter, standard deviation, 159, 160,
    164, 172, 176, 181, 185
Zones, 6, *see also* Inhibition zones